SD329

The Open University

SIGNALS and PERCEPTION: THE SCIENCE OF THE SENSES

5
6
7

The Open University, Walton Hall, Milton Keynes, MK7 6AA

First published 2003, Reprinted 2006.

Edited, designed and typeset by The Open University.

Printed in the United Kingdom at the University Press, Cambridge.

ISBN 07492 35780

This publication forms part of an Open University course, *SD329 Signals and Perception: the science of the senses*. The complete list of texts which make up this course can be found at the back. Details of this and other Open University courses can be obtained from the Call Centre, PO Box 724, The Open University, Milton Keynes MK7 6ZS, United Kingdom: tel. +44 (0)1908 653231, e-mail ces-gen@open.ac.uk

Alternatively, you may visit the Open University website at http://www.open.ac.uk where you can learn more about the wide range of courses and packs offered at all levels by The Open University.

To purchase this publication or other components of Open University courses, contact Open University Worldwide Ltd, Walton Hall, Milton Keynes MK7 6AA, United Kingdom: tel. +44 (0)1908 858785; fax +44 (0)1908 858787; e-mail ouwenq@open.ac.uk; website http://www.ouw.co.uk

1.2

The SD329 Course Team

Course Team Chair
David Roberts

Course Manager
Yvonne Ashmore

Course Team Assistant
Margaret Careford

Authors
Mandy Dyson (Block 3)

Jim Iley (Block 6)

Heather McLannahan (Blocks 2, 4 and 5)

Michael Mortimer (Block 2)

Peter Naish (Blocks 4 and 7)

Elizabeth Parvin (Blocks 3 and 4)

David Roberts (Block 1)

Editors
Gilly Riley

Val Russell

Indexer
Jean Macqueen

OU Graphic Design
Roger Courthold

Jenny Nockles

Andrew Whitehead

CD-ROM and Website Production
Jane Bromley

Eleanor Crabb

Patrina Law

Kaye Mitchell

Brian Richardson

Gary Tucknott

Library
Judy Thomas

Picture Research
Lydia Eaton

External Course Assessors
Professor George Mather (University of Sussex)

Professor John Mellerio (University of Westminster)

Consultants
Michael Greville-Harris (Block 4, University of Birmingham)

Krish Singh (Block 2, Aston University)

BBC
Jenny Walker

Nicola Birtwhistle

Julie Laing

Jane Roberts

Reader Authors
Jonathan Ashmore (University College London)

David Baguley (Addenbrooke's Hospital, Cambridge)

Stanley Bolanowski (Syracuse University)

James Bowmaker (University College London)

Peter Cahusac (University of Stirling)

Christopher Darwin (University of Sussex)

Andrew Derrington (University of Nottingham)

Robert Fettiplace (University of Wisconsin)

David Furness (Keele University)

Michael Greville-Harris (University of Birmingham)

Carole Hackney (Keele University)

Debbie Hall (Institute of Hearing Research, Nottingham)

Anya Hurlbert (University of Newcastle upon Tyne)

Tim Jacob (University of Cardiff)

Tyler Lorig (Washington and Lee University)

Ian Lyon (Consultant)

Don McFerran (Essex County Hospital)
Keith Meek (University of Cardiff)
Tim Meese (Aston University)
Julian Millar (Queen Mary, University of London)
Peter Naish (Open University)
Robin Orchardson (University of Glasgow)
Alan Palmer (Institute of Hearing Research, Nottingham)
Krish Singh (Aston University)
Charles Spence (University of Oxford)
Rollin Stott (DERA Centre for Human Sciences)
Steve Van Toller (University of Warwick)
Stephen Westland (University of Derby)

BLOCK FIVE

Touch and Pain

Contents

1 Introduction	**5**
1.1 Summary of Section 1	6
2 Touch	**7**
2.1 Properties of tactile mechanoreceptors	7
2.2 Somatosensory coding	17
2.3 Summary of Section 2	18
3 The perception of touch	**19**
4 Proprioception	**25**
4.1 Properties of proprioceptors	25
4.2 From neuron to brain: posture, balance and movement	28
4.3 Summary of Sections 3 and 4	29
5 Pain	**31**
5.1 Tissue damage	33
5.2 Properties of nociceptors	37
5.3 Transduction of noxious stimuli	41
5.4 Afferent pathways and the gate-control theory of pain	43
5.5 The control of pain	44
5.6 Summary of Section 5	48
Objectives for Block 5	**50**
Answers to questions	**51**
Acknowledgements	**190**
Glossary for Blocks 5, 6 and 7	**193**
Index for Blocks 5, 6 and 7	**201**

Introduction

1

When asked to name the human senses, most people would name the five obvious ones: vision, hearing, touch, smell and taste. Few people would include balance, pain, pressure, temperature, position and movement, the senses that together with touch comprise the somatic sensory system or 'somatosensation'. These, then, are the **general senses**; senses that are found scattered around the body. They are not evenly spread amongst the body organs (the **viscera**) and tissues but nor are their receptors to be found in discrete locations such as those of the **special senses** you have studied, the ear and the eye.

So what do we know about the senses that are grouped together as somatosensation?

❍ Spend a moment with your eyes closed in a quiet place and focus on your body. Try to be aware of the position of your body and the stimuli which it can detect. Then slowly move your arms until they make contact with another object, or another part of your body. What stimuli were you aware of?

● Almost certainly you were aware of touching when your arms or hands contacted another object and you may have been aware of some qualities possessed by the object touched. Was it hot or cold, smooth or rough? Before you moved your arms you would have been aware of the position of your body and the relation of the various parts, one to another. Were you sitting or standing, arms bent or stretched?

Unless you bumped into a sharp object or were bitten by an insect during this exercise, this is probably all you were aware of; and you might not even have been *consciously* aware that you knew about your body position. There are other aspects of our internal environment that also need to be monitored, so that they can be responded to and adjustments made. For example, we have receptors that are sensitive to blood pressure and others that are sensitive to its chemical composition. If you have ever taken instruction in yoga or meditation you may have attempted to become consciously aware of stimuli such as these.

Somatosensation, therefore, includes senses of which we are normally unaware. The stimuli are transduced by specialized neurons rather than by receptor cells such as the hair cells of the cochlea and the rods and cones of the retina. There are many different types of these neurons, and each specialized neuron is responsive to its own **adequate stimulus** (i.e. its own particular stimulus modality).

It is usual to subdivide somatosensation, and to discuss the sub-systems separately. There are a number of different ways to do this. In this block we focus on touch, proprioception, and on pain. Both touch and proprioception depend upon activity in a variety of mechanoreceptors, that is receptors that respond to mechanical stimulation. Unfortunately, the mechanism(s) by which mechanical stimuli lead to changes in membrane potential are not known in any detail so we are not including any physical details of this process here.

Pain is the perception that can arise from stimulation of **nociceptors**. These include receptors that are sensitive to tissue damage, so we will be looking at the biochemical events associated with tissue damage. However, nociceptors are unusual in that some respond to mechanical and thermal stimuli as well as to tissue damage. These are therefore described as **polymodal receptors**.

A sufficiently large blow to the head can generate a response in the optic nerve but it will be interpreted in the visual cortex as a visual signal (hence the expression 'seeing stars'). By contrast, when nociceptors are stimulated by a cut to the skin (tissue damage, therefore chemoreception) the perception differs from that generated by a blow to the same area of skin – even though the same nociceptors are being stimulated by each event. By the time you have completed this block you will have been introduced to the mechanism that is believed to mediate this particular phenomenon of perception.

Another category of somatosensation is the sense of temperature. This is sometimes discussed alongside nociception because the receptors are morphologically similar and the afferent pathways are the same for both senses. Furthermore it could be argued that we are not usually aware of the temperature of our surroundings or of objects in it, unless they are slightly uncomfortable. For the reasons given above you will find that there is mention of thermal detection in the Reader chapters on pain but no separate chapter on this sense.

Table 1.1 summarizes the way that we will group the receptors that are active in somatosensation. When you read Chapter 19, *The perception of touch*, you will notice that Peter Cahusac has discussed all of the sensory modalities listed in Table 1.1 whilst retaining the major focus on 'touch'. This demonstrates clearly that there is an external validity to our consideration of somatosensory systems alongside visual and auditory systems. Somatosensation is not a 'rag bag' classification.

Table 1.1 Sensory modalities and associated receptors.

Sensory modality	Receptor type	Location of receptors
touch (including pressure and vibration)	mechanoreceptor	skin and deep tissues
proprioception (including muscle stretch, tension and joint position)	mechanoreceptor	muscles, tendons and joints
temperature	thermoreceptors	skin, hypothalamus
nociception (tissue damage or disturbance)	various (including polymodal)	skin and various organs

Table 1.1 does not include the receptors responsible for monitoring the physiological state of the body because most authorities consider these separately from somatosensation and we are not going to discuss them further in this course. However, somatosensation is strictly defined as the sensations that are conveyed to the central nervous system (CNS) by peripheral sensory afferent neurons that have their cell bodies in the dorsal root ganglia (DRG) and so those receptors are technically part of the somatosensory system.

We shall begin our consideration of somatosensation with touch, before moving on to proprioception and then pain.

1.1 Summary of Section 1

Somatosensation, the bodily senses, includes a variety of senses that share the common feature of using a modified sensory neuron as their receptor. Somatosensation is usually divided into four sensory modalities: touch, proprioception, nociception and temperature, but these modalities combine to provide a unified perception of our bodily sensations and of those objects with which we physically interact.

Touch

2

Touch, sometimes also called **taction**, encompasses the range of sensation produced by the lightest of touches to pressure that is just short of painful. Physical contact with the skin is not required for this perception. The movement of the air can stimulate some of our receptors; for example when you blow on the back of your forearm you may feel hairs move. It is the receptors around the hair follicles (embedded in the skin) that are stimulated.

2.1 Properties of tactile mechanoreceptors

It has already been mentioned that there are many different types of mechanoreceptors embedded in the skin and they do not all respond to the same stimuli. Figure 2.18 of Block 2 showed some of these receptors and we reproduce that figure here to remind you of the different receptor characteristics (Figure 2.1). We will now look at them and their functions in more detail.

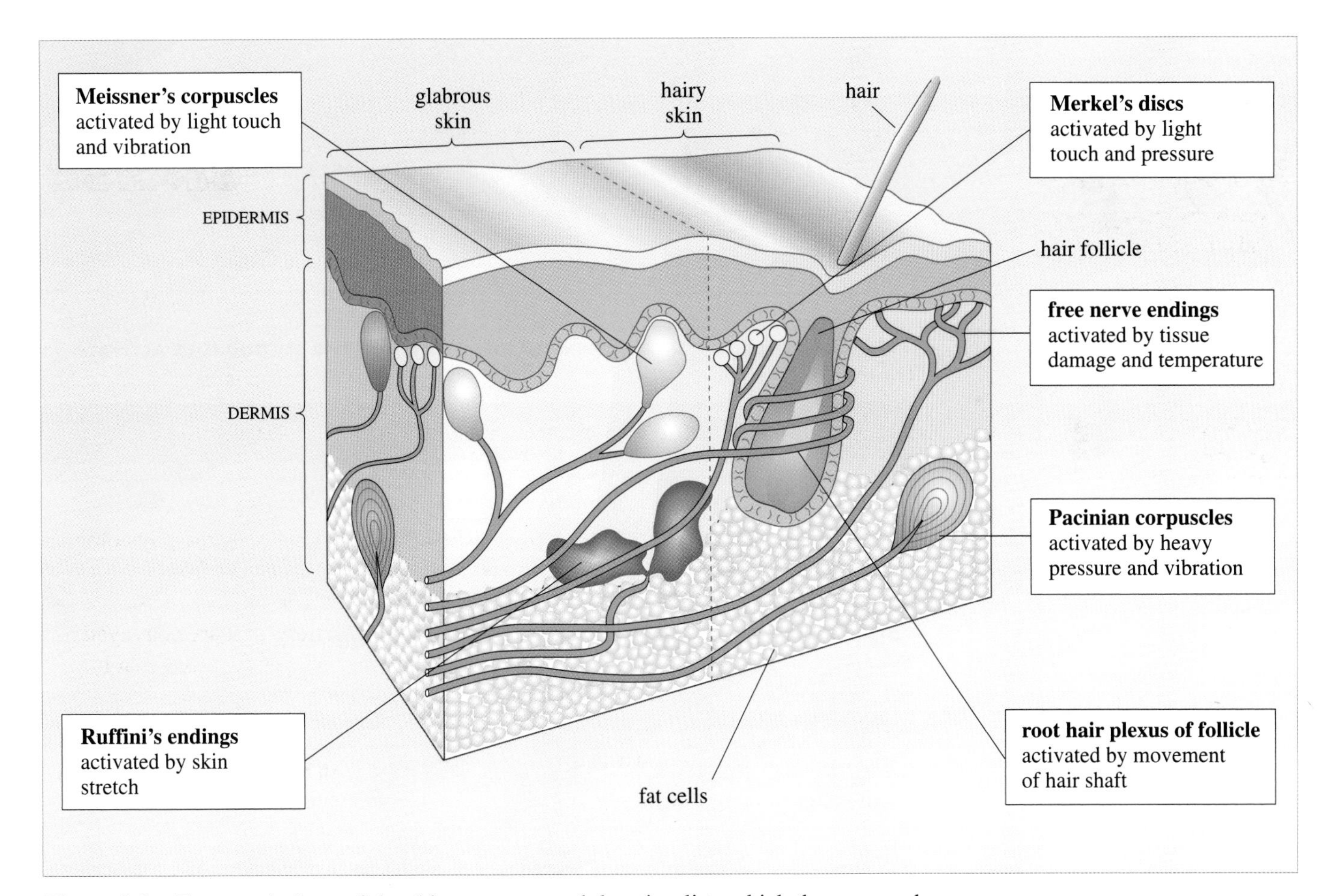

Figure 2.1 The morphology of the skin receptors and the stimuli to which they respond.

Figure 2.2 (overleaf) shows diagrammatically some peripheral receptive elements. Table 2.1 (overleaf) gives details of the adequate stimulus for different receptor types.

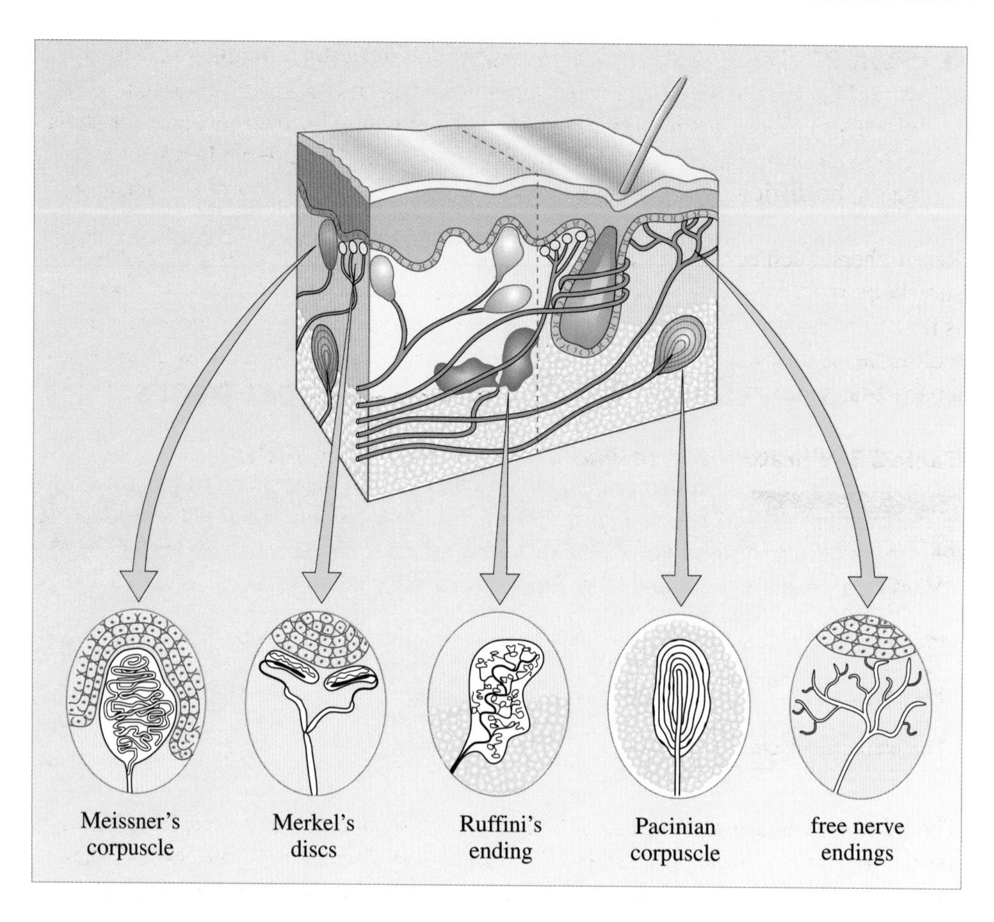

Figure 2.2 Some of the receptor types that are found in human skin.

Table 2.1 Receptor types involved in taction and the stimuli that activate them.

Receptor type		Adequate stimulus
Merkel's disc		pressure
Meissner's corpuscle		stroking
Pacinian corpuscle		vibration
Ruffini ending		skin stretch
hair follicle receptors	tylotrich	stroking
	hair-guard	stroking
	hair-down	light stroking
field receptors		skin stretch

Although Table 2.1 gives specific information about the adequate stimulus for each receptor type, the relationship between a stimulus, the receptor(s) that are responding and the sensation experienced is anything but clear cut.

❍ Can you think why it might be difficult to establish the relationship between the stimulus, the skin receptor and the sensation?

- Figure 2.2 shows the receptor types embedded in fleshy, vasculated tissue. It would be difficult to know which types of receptor(s) were responding to any given stimulus. Also it would be hard to isolate an individual receptor for study. When recording from a single axon within the peripheral nerve bundle it would, again, be difficult to know what type of terminal receptor was at the other end!

Researchers, such as the American, Stanley Bolanowski, (whose chapter you will shortly be reading), have devised ingenious ways around the problem. One example is that cooling the skin 'knocks out' some receptor types. Using psychophysical techniques it then becomes possible to identify the range of stimuli that gives rise to a particular sensation (Table 2.2) and to postulate the receptor type that is responding.

Table 2.2 Vibratory stimuli and the perceptions they elicit.

Receptor	Best frequencies	Perception
Merkel's disc	0.3–3 Hz	pressure
Meissner's corpuscle	3–40 Hz	flutter
Ruffini ending	15–400 Hz	buzzing
Pacinian corpuscle	10–500 Hz	vibration

The other way to approach the problem is to both stimulate and record from a single axon within the peripheral nerve bundle. In this way the response properties of the axon can be determined. This, with the psychophysical results and **histological** examination of the skin (i.e. the microscopic examination of the structures in the skin) gives us a picture of the way that the individual receptors detect particular aspects of a stimulus. Nevertheless we should not underestimate the difficulties of studying receptors that are scattered throughout the body. Most of the terminal receptive elements are tiny and in some areas they are densely packed. For example, the skin at the tip of your finger is innervated by around 300 sensory afferent fibres/cm^2. Each fibre terminal can branch as many as 25 times, each branch being, functionally, a receptor.

You might have noticed that the free nerve endings (Figure 2.2) are not included in Tables 2.1 and 2.2. These cell types are the most common class of skin receptors. They are particularly implicated in the detection of thermal stimuli and tissue damage, and so are not explicitly included when considering taction. However, we cannot completely ignore them because some of them are certainly responsive to light touch and pressure; for example, in the cornea of the eye they are the *only* mechanoreceptors to be found (Box 2.1 overleaf). So here they are definitely responsible for detecting and encoding information about tactile stimuli.

In Block 2, Figure 2.35 showed the anatomical basis upon which it would be possible for two closely situated stimulus sites to be discriminated. The receptors involved were drawn as branching lines for simplicity but from the previous paragraph it is clear that (except in the cornea) free nerve endings are not mediating this response. So which receptor type is responsible for this discriminatory ability?

- ❍ Figure 2.1 tells you which receptors are sensitive to a light touch. Which are they?
- ● Meissner's corpuscles and Merkel's discs.

Box 2.1 The mechano-nociceptors of the cornea

Mechano-nociceptors are found throughout the body. In normal, undamaged tissue they are activated only by intense mechanical stimulation, such as pinching or heavy pressure. But, the situation in the eye is different. Almost certainly, you will have experienced that contact between the cornea and the most minuscule of objects is instantly detected. And a good thing too, because it results in the removal of whatever it was in your eye long before any damage befalls the cornea! This protective function of the tactile mechano-nociceptors of the cornea also means that it takes practice to be able to insert a contact lens. In Chapter 9 of the Reader it was stated that the tear film assists in smoothing out any irregularities on the front surface of the cornea. It is the tear film that holds the lens in place. Once in place the lens cancels the refraction that usually takes place at the anterior surface of the cornea because the tear film has a refractive index almost equal to that of the cornea. Thus the contact lens' anterior surface becomes the significant refractive surface in the eye's optical system. Contact lenses are particularly beneficial for people who have an irregularly shaped cornea, for example those individuals with a bulging, odd-shaped cornea in a condition known as *keratoconus,* for whom spectacles cannot give a satisfactory solution. So it is worth the initial difficulty of getting the hang of using these lenses. Once inserted, the wearer is not troubled by the 'feel' of the lens.

The areas with high tactile acuity (such as fingertips and tongue) must have numerous receptors, each with a small receptive field. So do different receptors have differently sized receptive fields dependent upon where they are located, or do some receptor types have small receptive fields and other receptor types have large receptive fields? Valbo and Johansson, working in America in 1978, discovered that there were two kinds of axons within the peripheral nerve bundles from the hands that had small receptive fields. The receptors involved are Meissner's corpuscles and Merkel's discs. Both of these receptor types are found close to the surface of the skin, just below the epidermis, though some Merkel's discs are in the epidermis (Figure 2.1).

There are however other, more deeply located receptors in the hands such as Pacinian corpuscles and these have large receptive fields. It seems then, that each type of receptor has a characteristically sized receptive field, 'small' or 'large', although the actual dimensions of a 'small' receptive field vary depending on their precise location as you have already discovered by doing the two-point discrimination test in Block 1. But why should there be two types of receptors with small receptive fields? Wouldn't just one do the job?

❍ What differences in function are shown for Merkel's discs and Meissner's corpuscles in Table 2.2 and also in Figure 2.1?

● These two receptors signal the presence of slightly different stimuli. Both receptors will respond to light touch but only Meissner's corpuscles respond to vibratory stimuli (Table 2.2) and apparently only Merkel's discs respond to pressure.

Table 2.3 shows how the responses recorded from the axons of these and other receptor types differ when the skin receives a constant pressure stimulus.

Table 2.3 Receptor types found in hairy and glabrous (non-hairy) skin, classified by rate of adaptation and location.

	Hairy skin	Glabrous skin	
Adaptation to constant pressure stimulus:			Location in skin:
very rapid	Pacinian corpuscle	Pacinian corpuscle	*deep*
rapid	hair-follicle receptors	Meissner's corpuscle	*superficial*
slow	Merkel's discs	Merkel's discs	*superficial*
slow	Ruffini endings	Ruffini endings	*deep*

A feature of the different mechanoreceptors shown in Table 2.3 (and briefly mentioned in Block 2) is that some of them adapt rapidly to constant pressure. Try blowing gently on the back of your hand: move your hand slowly around whilst trying to keep a steady current of air flowing through your lips. Now think about the part of your body that is supporting you, your feet (if you are standing) or your bottom (if you are sitting).

❍ Which part of your body was experiencing the greater pressure?

● Your feet (or bottom) would be experiencing more pressure than the back of your hand.

❍ Which stimulus was the more noticeable?

● Many people find blowing on the back of the hand produces the more noticeable stimulation.

❍ Why might this be so?

● It was mentioned in Block 2 that the CNS is particularly alert to *changes* in stimulation. However hard you try it is difficult to produce a completely steady air flow when you blow! Also you were asked to move your hand which ensured that the stimuli were constantly changing. A second factor, also mentioned in Block 2, is that some mechanoreceptors adapt to a stimulus very rapidly. You were told that this was why you didn't consciously feel your clothes all the while. Equally, it accounts for your not being aware of pressure on your bottom all the time you remain seated.

It is not necessary or desirable for us to be continuously aware of unchanging stimulation (such as the presence of our clothing). The fact that we are unaware of unchanging stimuli, such as the presence of clothing is due to these rapidly adapting receptors. Both Meissner's corpuscles and hair follicle receptors only fire at the onset or termination of a stimulus (Figure 2.3). They are therefore suitable for signalling *dynamic* aspects, such as when a stimulus is changing; additionally, they can provide information about the rate of change of the stimulus.

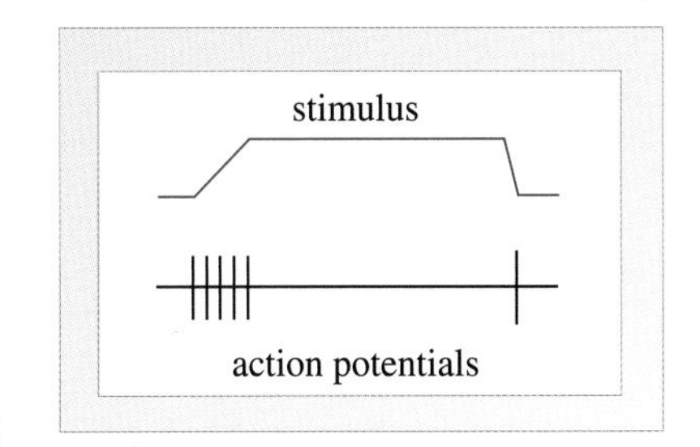

Figure 2.3 Firing of rapidly adapting receptors occurs at the onset and at the termination of a stimulus.

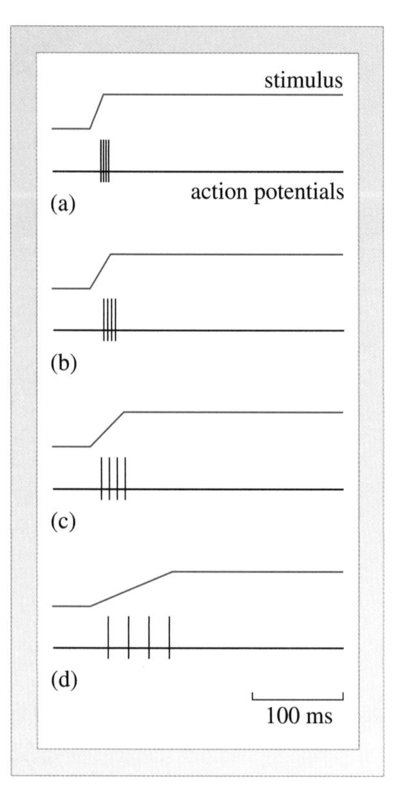

Figure 2.4 Responses from a rapidly adapting receptor, e.g. a Meissner's corpuscle, to different rates of application of a 'ramp-and-hold' stimulus. The rate of application is shown by the slope of the 'ramp', which is fastest in trace (a) and slowest in trace (d). The number of action potentials generated is the same in each case, but the frequency of firing is highest with the most rapidly applied stimulus (a).

Figure 2.4 shows how the firing rate of a rapidly adapting receptor increases as the speed of application of a **ramp-and-hold** stimulus increases. Thus rapidly adapting receptors can function as velocity detectors.

Slowly adapting receptors (e.g. Ruffini endings and Merkel's discs) can also serve this function, but additionally since they generally maintain their firing during the stimulus, they can also convey information about the *static* aspects of stimulus intensity and function as intensity detectors (Figure 2.5).

How does this relate to how we feel objects? Consider your fingertips: they have two types of receptors close to the surface of the skin, each doing a slightly different job. Meissner's corpuscles alert you to a single skin contact and also, as just explained, to changes in pressure (i.e. to the dynamic aspects of the stimulus) such as might occur if a rough surface, sandpaper for example, were to be rubbed across your skin. In other words, they give information about the texture of the material. But suppose that a tiny thorn stuck into your skin. Once lodged there the pressure would not alter. (Let us assume this is not actually painful.) Figure 2.4 shows that after an initial burst of activity the Meissner's corpuscle receptors rapidly fall silent.

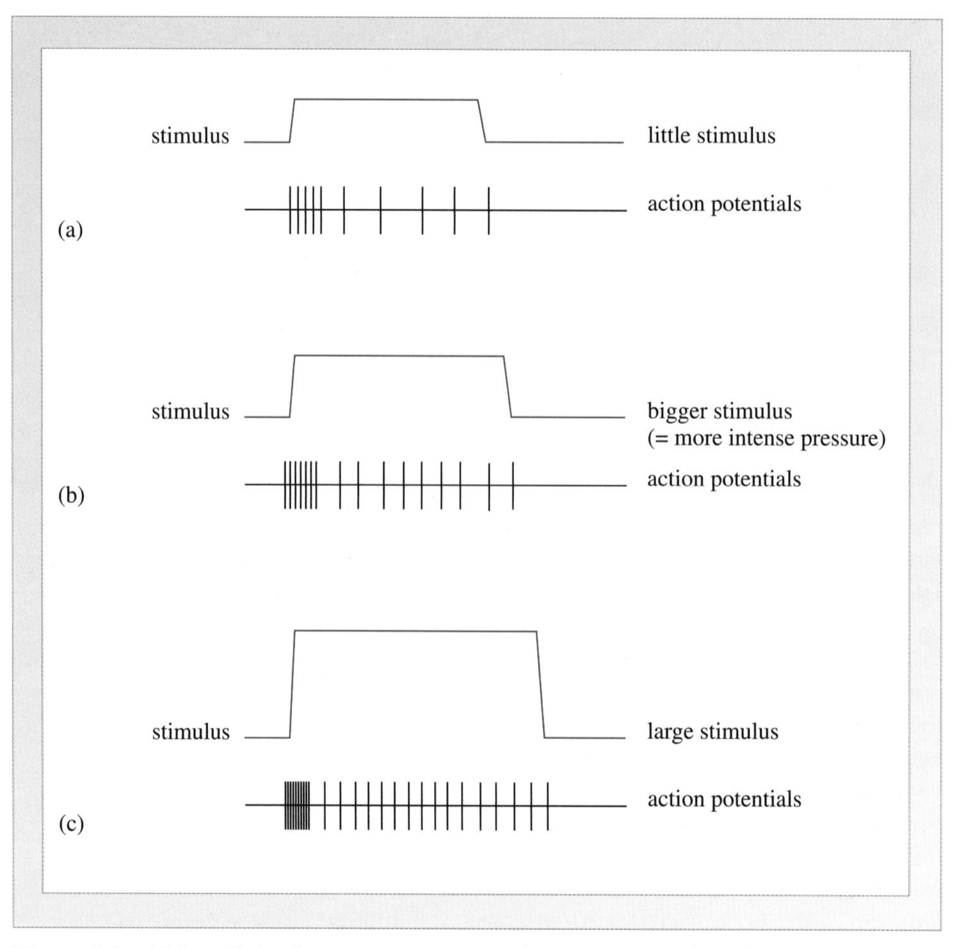

Figure 2.5 Firing of slowly adapting receptors in response to stimuli of different intensities: (a) a little stimulus results in the lowest frequency of action potentials both in the initial burst of electrical activity and the following period of maintained response; (b) a larger stimulus represents more intense pressure on the skin surface and the frequency of action potentials increases; (c) a large stimulus results in the greatest frequency of action potential spikes.

However, Merkel's discs would continue to fire for longer (Figure 2.5), though even they adapt and cease firing eventually. So while Merkel's discs remain active you continue to be aware of the presence of the thorn and the small receptive fields give you the precise location. If you don't manage to remove any or all of the thorn you will soon be able to ignore it as the receptors adapt to the unaltered pressure imposed.

With these two receptor types capable of responding to a range of stimulus attributes, what purpose can be served by having the more deeply located Pacinian corpuscles and Ruffini endings, with their large receptive fields? Obviously, because of its large receptive field, the Pacinian corpuscle cannot 'tell' the brain exactly where the stimulus is. The Pacinian corpuscle is a rapidly adapting receptor, tending to respond with only one or two action potentials when a stimulus is applied or removed. It is not very sensitive to the rate of application or the magnitude of the stimulus. For example, in Figures 2.6a and 2.6c the responses to two different levels of stimulation are the same. However, the Pacinian corpuscle is very sensitive to oscillating stimuli such as those shown in Figures 2.6b and 2.6d. Here, each cycle of the stimulus generates an action potential. These oscillating stimuli are felt by us as vibration, and the Pacinian corpuscle might be described as a vibration detector. Table 2.2 shows Ruffini endings to have very similar properties.

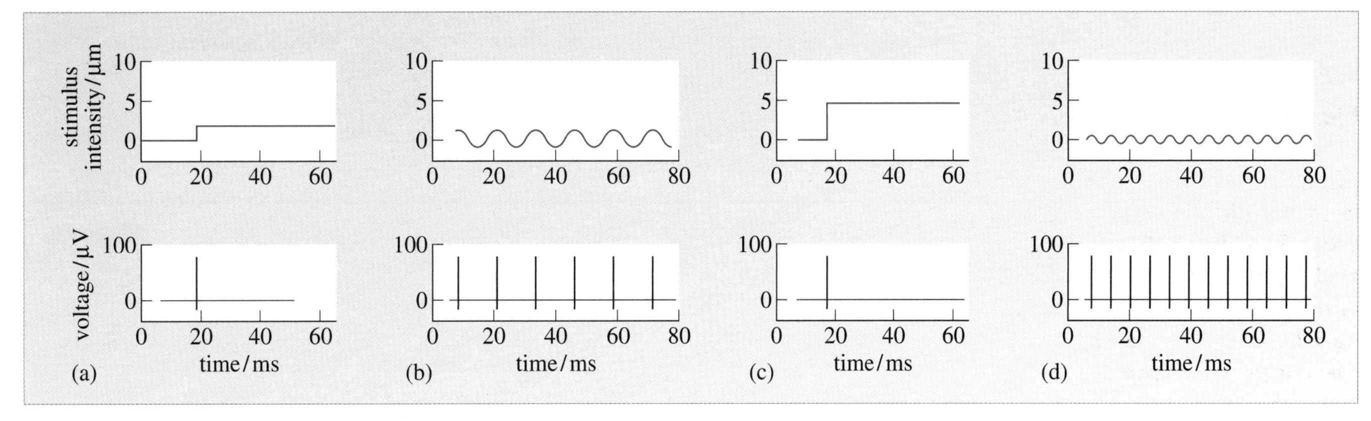

Figure 2.6 Response characteristics of Pacinian corpuscles. The upper panels show the form of the stimulus applied; the lower panels display the action potentials generated by these stimuli. In (a) and (c), single mechanical stimuli of different intensities are applied. In (b) and (d), the mechanical deformation oscillates evenly at two different frequencies.

From the table we can surmise that neither Ruffini endings nor Pacinian corpuscles are much use at detecting a single event such as the thorn piercing your skin (because the generation of one action potential is not going to result in an event being detected at the cortical level). A stimulus that occurs once only is of vanishingly low frequency! They are, however, very sensitive to high frequencies (i.e. vibrations). The usefulness of being sensitive to a high frequency isn't immediately obvious, but it is what alerts you to the fact that something is slipping from your grasp (or should you be climbing a tree it's the first indication that you are slipping!) This information is pretty useful to tool-using animals like us!

We can also detect our skin stretching or being stretched. Try pulling a few grotesque faces to experience the feel of skin stretch. Ruffini endings and field receptors are believed to be the major responding sensory receptors in this situation.

In Block 3 we explained the basic principles of psychophysics and applied them to studies on hearing.

❍ In what way do the receptors for sound resemble those used in touch?

● Both senses employ mechanoreceptors.

❍ What is the main aim of psychophysics?

● It attempts to quantify the relationship between a stimulus and the sensation it evokes (Block 3, Section 5).

❍ What are the three general approaches used?

● (1) Measurement of the detection threshold (i.e. the smallest value of a stimulus that can be detected); (2) measurement of the smallest difference between two stimuli that can be discriminated; (3) scaling procedures (this involves describing a stimulus).

Each of the above approaches has been used in the study of touch to reveal details of the properties of tactile mechanoreceptors.

❍ Which of the three approaches has been used in order to plot the graphs shown in Figure 2.7?

● Approach 1. The graph shows the detection threshold for various stimuli. The stimuli used range from a single touch to vibratory stimuli with frequencies above 1000 Hz.

Just as with the other senses you have studied, the graph shows that the intensity at which we detect a stimulus varies according to its frequency. We have noted that Pacinian corpuscles are not very sensitive to a single stimulus, whatever its intensity, and this is clearly shown in Figure 2.7 (there is no response on the left-hand side of the graph). Figure 2.7 also shows how the properties of Meissner's corpuscles account for almost all of our ability to detect a single touch.

❍ Meissner's corpuscle receptors are responsible for detecting the probe (Figure 2.7). Will they signal increased pressure by firing more rapidly?

● No! Meissner's corpuscle receptors are rapidly adapting receptors (Table 2.3) and Figure 2.3 shows that the rate at which they fire codes for the rate of application of the stimulus.

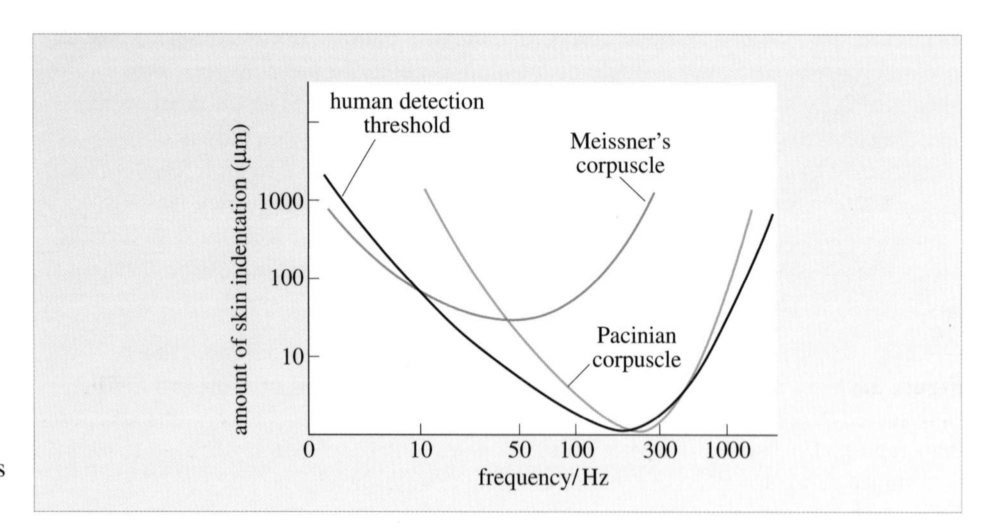

Figure 2.7 Graph of detection threshold for skin indentation (i.e. a measure of the intensity of stimulation) at different frequencies (i.e. different vibratory stimuli).

❍ How do you suppose they might code for intensity?

● Intensity can be coded for by the number of receptors responding (as was first suggested in Block 2, Section 2.2.2).

(It is also the case for vibratory stimuli that as the intensity of the stimulation increases more receptors respond.)

❍ Is this the only way that an increase in pressure is coded?

● No. Merkel's discs fire more rapidly when the pressure is greater (Figure 2.5).

Figure 2.7 shows data gathered from one area of skin only. If you now study Figure 2.8a you will see that the threshold for the detection of a single stimulus varies across the body surface.

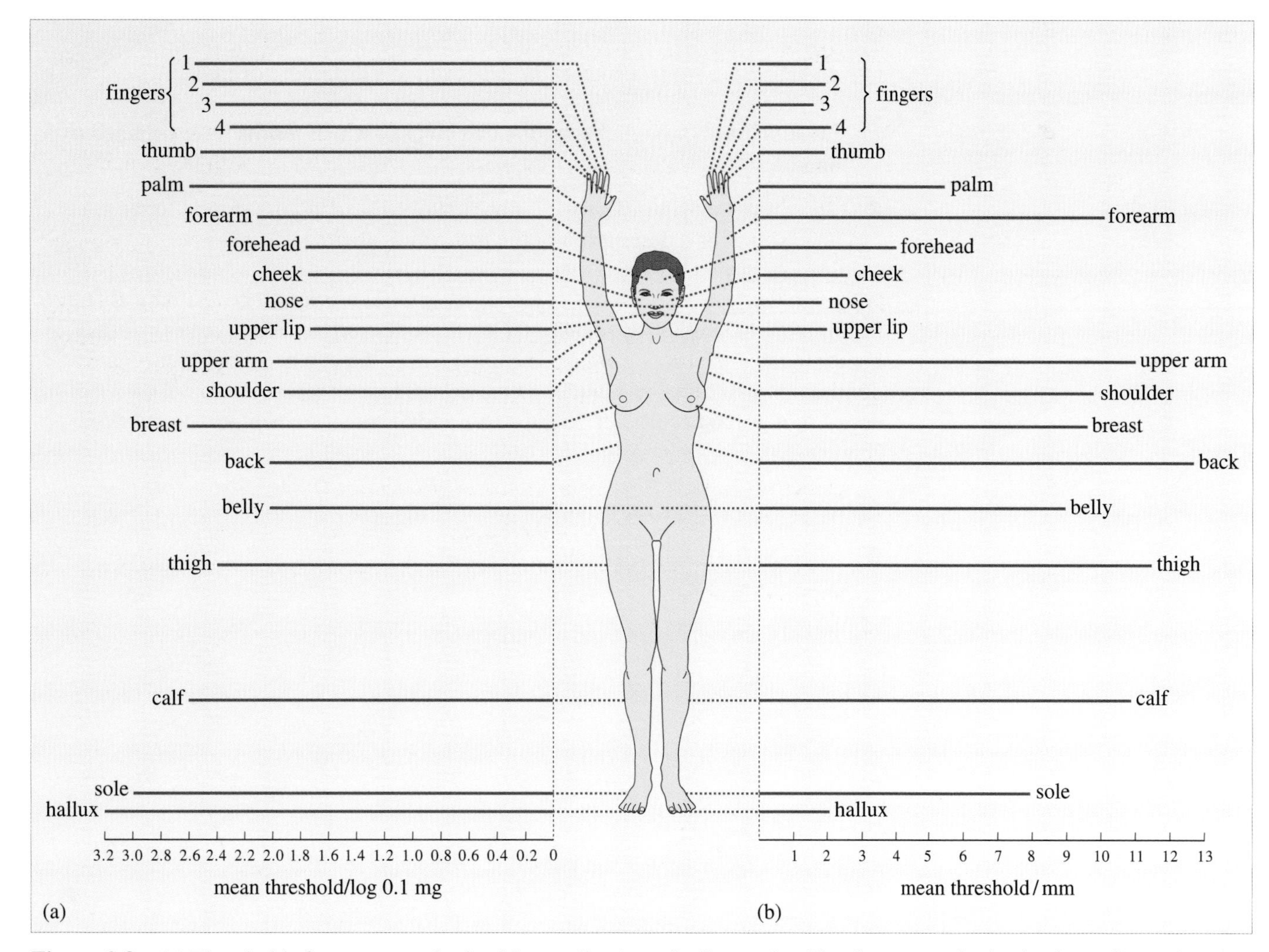

Figure 2.8 (a) Thresholds for pressure obtained by application of a fine probe. The data were obtained using a fine nylon fibre (calibrated in mg) as a probe. The intensity of the pressure applied to the skin surface can be accurately calibrated. The lines represent the amount of pressure that can be applied before the probe is detected. (b) Point localization thresholds determined using a single-touch stimulus. The hallux is the big toe.

- ❍ From Figure 2.8a, which parts of the body are (a) the most sensitive and (b) the least sensitive to pressure.
- ● (a) The face is the most sensitive and (b) the big toe is the least sensitive. It takes almost twice as much pressure to feel something on the toes than to feel it on your face.

Whilst the big toe may not be very sensitive to pressure, Figure 2.8b shows that it is rather good at localizing any stimulus that it does detect, as you would probably know from experiencing a small stone in your shoe. Thus the relative sensitivity of different parts of the body to pressure does not reflect their relative ability to accurately locate a stimulus (Figure 2.8).

- ❍ Which of the three approaches was used to obtain the data for Figure 2.8b?
- ● Approach 2. Point localization thresholds are determined using a single touch stimulus.

This involves the determination of the minimum distance that the probe has to be moved before the subject detects the sensation at a different position.

- ❍ Looking back through the section can you say which results were obtained using approach 3?
- ● The results shown in Table 2.2 were obtained by applying stimuli to the skin and asking for a description of the perception elicited.

Although there are many types of tactile receptors, they have one anatomical feature in common. As previously mentioned, all these sensory receptors have their cell bodies in the dorsal root ganglia. The axon leaves the cell body and immediately divides into two fibres or branches. One axonal branch enters the spinal cord via the dorsal root portion of the peripheral nerve and the other branch goes to the skin (Figure 2.9).

- ❍ What typical neuronal feature is missing in these neurons?
- ● They do not have dendrites.

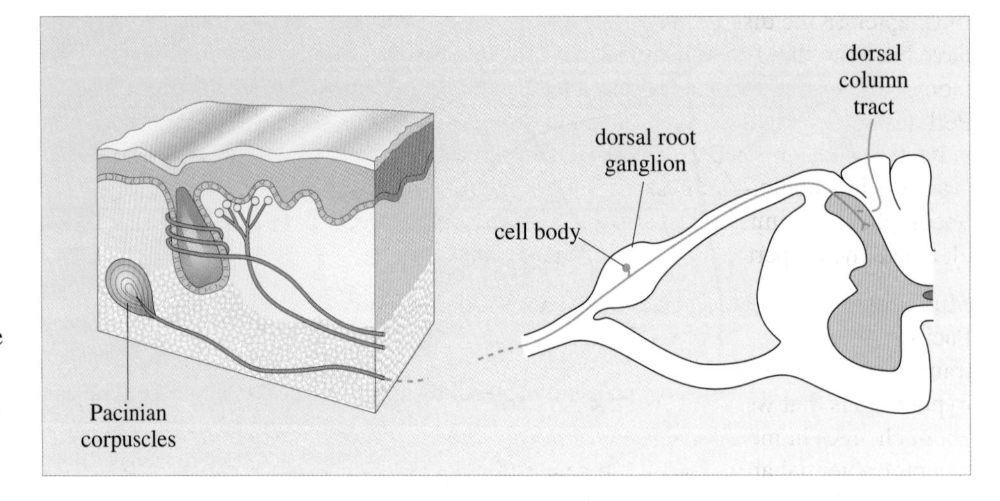

Figure 2.9 Diagram showing the location of the cell body of the somatosensory mechanoreceptors, e.g. Pacinian corpuscles, and the bifurcating axonal processes.

By now you will be aware that many neurons do not conform to the 'classic picture' given in Block 2, Figure 2.6. Many of the receptors in the skin have non-neuronal connective tissue elements closely associated with, or surrounding the actual ending of the axonal branch in the skin. Other receptors have no associated connective tissue elements, and are described as 'free nerve endings'. Transduction occurs at the axonal tips although we do not yet know how the mechanical stimulus is transduced. Does it physically distort the plasma membrane thereby opening ion channels (somewhat similar to the mechanism that activates the ear's hair cells) or is there some chemical mechanism involved (as with the rods and cones of the eye)? We cannot yet answer this question with complete certainty but where it has been possible to make recordings we do know that it is at the tip of the axon that a receptor (or generator) potential is first detected. And it seems that there must be voltage-gated Na^+ channels located just before the axon becomes myelinated, because this is where the action potential is first generated.

Although they are not believed to be directly party to the transduction mechanism, the non-neuronal elements of skin receptors do seem to mediate between the stimulus and the receptor's response. For example, an intact Pacinian corpuscle does not continue to respond to steady pressure, but it will do so if its corpuscle has been dissected away from the axonal tip. In these artificial circumstances the receptor potential is maintained but the afferent fibres response still adapts (i.e. it does not continue to fire).

Elucidating the properties of receptors has been (and still is) a piecemeal process and we know more about some types than others, partly because of the technical difficulties of isolating receptors and partly because most research efforts have concentrated upon the fine discriminative abilities of our fingertips. There are other specialized receptors that are not found in the hands and we have not considered them at all in this account. There are also problems of interpreting results gained from study of individual receptors' responses to simple stimuli when these receptors *in vivo* work collaboratively to provide the sensations we call touch.

2.2 Somatosensory coding

You should now read Chapter 18 of the Reader, *Somatosensory coding* by Stanley Bolanowski.

In Chapter 18 the discussion of afferent pathways and receptor morphology should have been familiar to you from Block 2. It is interesting that a range of stimuli (some not very sophisticated!) are used to investigate the properties of the receptors. Perhaps the use of items such as sandpaper and wool did not surprise you because it is items such as these that are significant in our lives. How often do we experience 'variable indentation of the skin by a probe'? Yet researchers need to use uncomplicated stimuli like a single nylon fibre probe as tools in their attempt to elucidate the properties of these mechanoreceptors.

Much of the work linking receptor anatomy and molecular events has used the Pacinian corpuscle – a relatively large receptor. But, the molecular basis of transduction is still not known. There is no evidence for it being chemical, so the hypothesis is that we will find stretch sensitive ion channels on filopodia that bring about changes in membrane potential – the receptor potential. Stimulus intensity and receptor potential are linearly related at low amplitude only. In other words, for larger stimuli equal increases in intensity do not manifest as regular increases in the size of the receptor potential because the response saturates (see Block 2, Figure 2.19). However, for a small range of stimulus variation, as the receptor potential increases, the axonal firing rate increases. But a steadily held stimulus (i.e. the 'hold' bit of the 'ramp-and-hold' stimulus, as shown in Figure 2.4a) will result in a fading of action potentials in all axons, though some show this response very rapidly (FA fibres) and others more slowly (SA fibres). Figure 7 of Chapter 18 summarizes the properties of the four skin receptors found in the hand.

Activity 2.1 Touch

At this point you may like to study the touch section of the 'Touch and Pain' sequence on *The Senses* CD-ROM. Further instructions are given in the Block 5 *Study File*.

Question 2.1

Draw the three routes for skin receptor afferents entering the spinal cord.

Question 2.2

Both Meissner's corpuscles and Pacinian corpuscles are rapidly adapting receptors. Use Figure 2.7 and Figure 7 in Chapter 18 of the Reader to refresh your knowledge of the differences in their other properties. Figure 7 in Reader Chapter 18 shows innervation density and receptive field sizes of the two types of sensory afferents. How would this relate to differences in the function of Meissner's corpuscles and Pacinian corpuscles?

2.3 Summary of Section 2

A variety of mechanoreceptor types subserve taction. Each one differs in structure and in its adequate stimulus. They are typically found at different depths below the skin surface but hairy skin has hair-follicle receptors in place of the Meissner's corpuscles found in glabrous (non-hairy) skin. The Pacinian corpuscle is the largest of these receptors and has been the easiest one to study. The transduction mechanism is not known but it is thought that there are stretch sensitive ion channels on the filopodia and that these are responsible for allowing ion movement across the most distal portions of the neurite.

Different receptors have differently sized receptive fields and show different rates of adaptation to a constant pressure stimulus. Receptors also vary in their response to oscillating (vibrating) stimuli.

Whatever class of stimulus is used, the intensity at which it is first detected can be accounted for by the detection thresholds of two receptor types: the Pacinian corpuscle and Meissner's corpuscle (both rapidly adapting receptors). It is also noteworthy that the threshold for the detection of a stimulus varies across the body surface but does not co-vary with the ability to localize a stimulus. Isolating the responses of the different receptors is not easy and in everyday life most stimuli cause more than one receptor type to respond. All the receptors are modified sensory neurons and enter the dorsal horn of the spinal cord. There are two ascending pathways, both of which project via the thalamus to the somatosensory cortex (parietal lobe) as well as a local circuit that provides an input to the motor system at the spinal level.

The perception of touch

3

The presence of a variety of receptors emphasizes that touch is not a unitary sensation. As you might expect by now, perceptual experiences that arise from tactile stimulation are not generally mediated by a single class of receptor but by a medley of sensory input that is interpreted as a unitary experience by somatosensory cortical processing. You will notice many similarities to processes you have met already. As with previous sensory modalities, we investigate the properties of the individual receptors and pathways to enable us to have a better understanding of the raw material from which the percept can be extracted. These investigations involve the experimenter applying different kinds of tactile stimuli to the skin. But we know that we *use* touch to give us information about the environment. We don't wait for stimuli to bump into us, we actively seek to touch objects. Why else would museums and exhibitions have to put up notices saying 'Do not touch' if it were not for the fact that this is such a basic way of exploring new objects. And as you will read in Peter Cahusac's chapter, *The perception of touch*, the ways of judging the quality of an item depend more often on touch than on sight. Would you ever select an item of clothing without first feeling the material? In fact there are so many ways that we feel an object that the psychologists Susan Lederman and Roberta Klatzky have produced a descriptive taxonomy of touch.

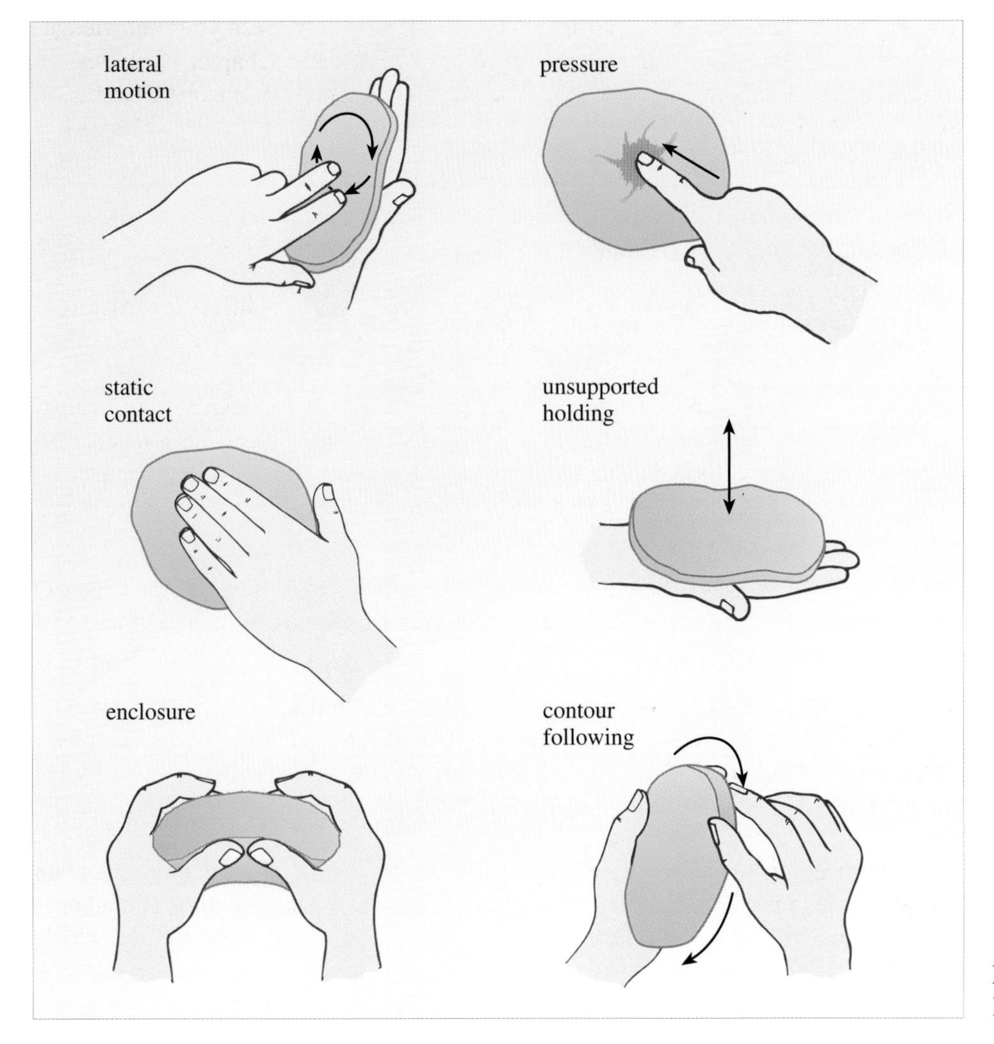

Figure 3.1 Lederman and Klatzky's taxonomy of touch.

Figure 3.1 shows some of the exploratory movements made by blindfolded subjects to identify objects. The hands work together to acquire information about the object. The 'lateral motion' is used to obtain surface texture information (roughness). 'Pressure' is applied by one hand to the object held by the other hand (this operation can also be done with one hand) to determine hardness. 'Static contact' is used to determine the temperature of the object or its thermal conductivity (see Box 3.1), and so give clues to the material used in the object's construction. 'Unsupported holding' consists of **hefting** the object up and down, and is used to determine the weight of the object. 'Enclosure' requires the fingers to mold around the shape of the object to obtain shape and volume information. 'Contour following' is used to obtain exact shape information. Lederman and Klatzky found that recognition of common objects by active touch was extremely accurate and fast. Blindfolded subjects manipulating common objects such as a golf ball, scissors and light bulb usually took less than 5 seconds for correct identification.

Box 3.1 Thermoreceptors and the use of temperature in touch

Touching an object (static contact in Figure 3.1) can give us quite a lot of information about the material(s) from which it has been made, based on its thermal conductivity. For example, metals are good conductors of heat so picking up a metal gardening implement on a cold day is a disturbingly unpleasant experience as the warmth from your hand is conducted away by the metal. It is an experience that you can avoid by wearing woollen gloves because wool is a poor conductor of heat, especially so if it has a loose weave thereby trapping pockets of air, an even worse conductor.

Thermoreceptors are of primary importance in the regulation of body temperature. Mostly this takes place without our being consciously aware of the physiological changes that are made. When we do become aware we alter our behaviour. Putting on more clothes or lighting a fire if the ambient temperature is low, and wearing gloves if we need to grasp metal tools.

Thermoreceptors are sensitive to small changes in skin temperature. There are two types, which are distinguished on the basis of their responses to skin temperature changes in the range 30–42 °C (Figure 3.2). 'Cold' receptors increase their rate of firing in response to a fall in skin temperature over this range; below 25 °C, their firing rate decreases. 'Warm' receptors increase their rate of firing when skin temperature increases in the range 30–45 °C, but have a sudden 'cut-off' at temperatures greater than 45 °C.

As with so many other receptors, a rapid change in temperature is a particularly strong stimulus. Temperature receptors have a steady background firing rate when the body surface is at 34 °C. Touching a hot or cold object will give an almost instantaneous increase in the firing rate of the hot (or cold) receptor.

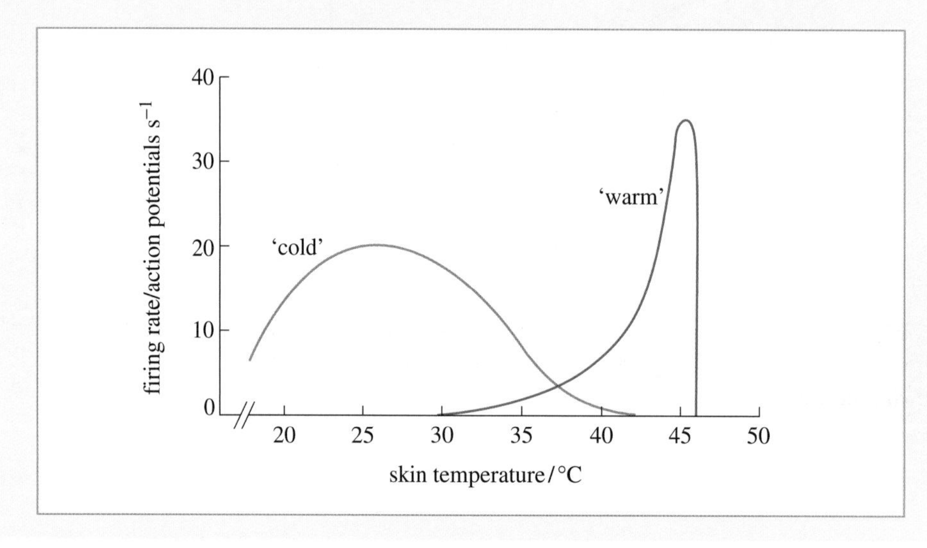

Figure 3.2 The relationship between action potential discharge and skin temperature in 'warm' (red line) and 'cold' (blue line) thermoreceptors.

When we use active touch we must know where the various parts of our body doing the touching (usually our fingers) lie relative to one another. Our index finger lies next to our thumb and if we pick up a pen between these two digits the pen will stimulate sensory receptors on the inner surfaces of both finger and thumb. If we did all this with our eyes closed then the sensory afferent information would inform us that we had been successful in this manoeuvre. We expect these anatomical relationships between the various parts of our body to remain constant – why should they not? – and we use this knowledge when we interpret incoming signals from the skin. The following simple activity should convince you of this.

Activity 3.1 Aristotle's illusion

Cross your index finger and your middle finger. Place a pencil on the V formed by their intersection (see Figure 3.3.) Now close your eyes and gently slide the pencil between the two fingers, touching both of them. You should experience a perception of being touched by two pencils.

This is known as Aristotle's illusion. (He is alleged to have first observed this using a small bead – the mind boggles at what he was doing at the time!) The explanation for the illusion is that the pencil (or bead) is rubbing the two *outside* surfaces of the pair of fingers and it would normally only be possible to do this simultaneously by using two pencils. Once again, as in previous blocks, we are noting the importance of expectation for our perceptions of the world around us.

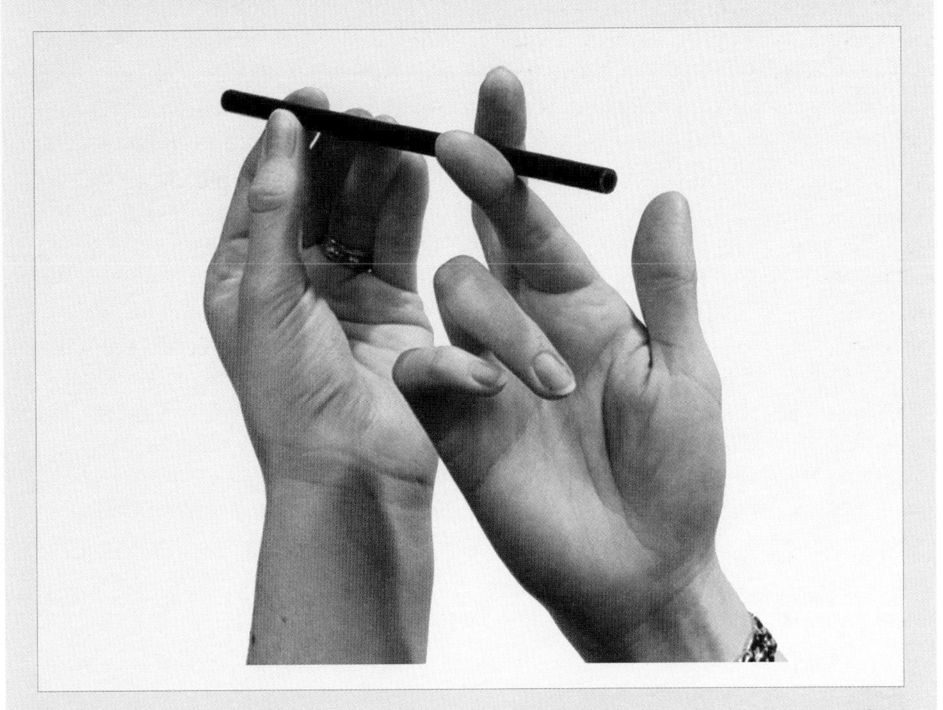

Figure 3.3 A demonstration of Aristotle's illusion. Note that it is important for the pencil to be touching the two outer surfaces of the crossed fingers to create this illusion successfully.

Box 3.2 Social aspects of touch

We would be deviating too far from the themes of this course to give too much attention to this issue but touch is so important to our social lives that it is a shame to ignore it completely. The everyday usage of 'feeling' to describe our basic emotions and our sense of how things are, in phrases such as 'you know how I feel about you' and 'my feeling is that this is the right decision' contrasts with, 'Yes, I see' when referring to understanding and knowledge. And this reflects our belief that feeling is a more primitive sense than seeing (see also Reader Chapter 20, *Proprioception*).

It is certainly the more important sense for establishing and maintaining social bonds. Right from the first moments of life when the most sensitive part of the baby's skin (whatever the advertisements say!) grasps the nipple to the moment of death when instinctively the living clasp the dying – even if in a sanitized, western hospital, it is only the hand that is held – it is touch that provides the all important contact between individuals (Figures 3.4 and 3.5).

The handshake and shaking on a deal for more formal relationships, and the less formal but stereotyped kiss on each cheek, both involve the most sensitive parts of the skin and are not common to all cultures. But even Eastern cultures would not deny the potency of such contact. Indeed, one important philosophical issue for the followers of K'ung Fu-tzu (Confucius *c.* 551–479 BC) was whether a man should allow his sister-in-law to drown if the only way to save her was to extend a helping hand (literally). The reason was that it might be preferable for her to die rather than to touch the hand of a man to whom she was not married and hence to live in shame.

The importance of touching and being touched for intimate relationships is too well known to need rehearsal here. An interesting insight into its importance in everyday life was gained from an experiment carried out at a library some years ago. After the librarians checked out the borrowers' books they made hand-to-hand contact in returning books to some of the borrowers. On leaving the library borrowers were asked for their opinions on a number of the library's facilities. Overall, the borrowers whose hands had been touched not only rated library facilities more highly, they were also more likely to judge that the librarian had smiled at them. The librarians had been instructed to smile at no-one.

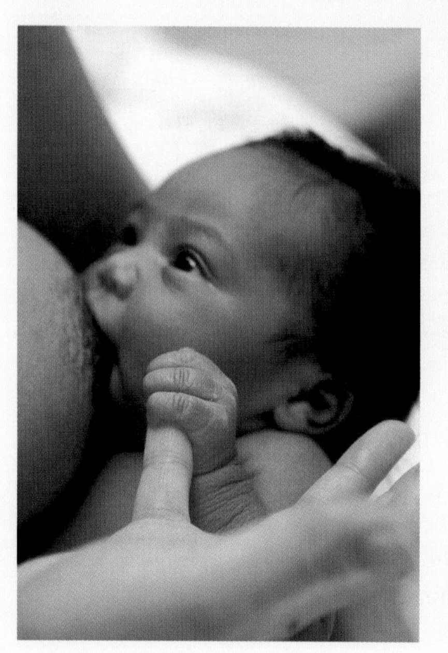

Figure 3.4 The importance of touch: the close bond between mother and breastfeeding baby.

Figure 3.5 Framed colour photograph of a mother holding her dead baby, taken *c.* 1857. The child most probably died from measles although chicken pox could be the cause of death here as the lesions on the face could have been caused by either disease. Memorial photographs of this type were common in nineteenth-century America. It is believed that these photographs helped the parents to move through the grieving process. Deaths from measles or chicken pox are rare now because of widespread vaccination programmes.

Now read Chapter 19 of the Reader, *The perception of touch*, and find out what Peter Cahusac has to say about the way we use touch.

Peter Cahusac's chapter shows us the extent to which we use active touch in our daily lives and how we integrate information across many senses at the cortical level. The interplay between taction and proprioception was strongly emphasized as was the fact that it is when things go wrong, whether through disease or illusion, that we often gain valuable insights into the working of the body and brain.

Check your understanding of two important issues raised in Chapter 19 by answering the following questions.

Question 3.1

It would appear that certain areas of the cerebral cortex are specialized for the analysis of specific sensory input (e.g. striate (occipital) cortex for vision, auditory cortex for sound). What evidence is presented in Chapter 19 that these functional areas might retain plasticity?

Question 3.2

Is there any evidence of crossmodal processing in cortical areas?

Question 3.3

In Chapter 19 of the Reader you were introduced to two terms: *body schema*, which means knowing where all the bits of your body are at any one moment (the continuous, subconscious update on where you are), and *body image*, a more familiar term, meaning an awareness of our physical attributes (for example, that we are tall and good-looking (or not!)). At the end of Chapter 19 your car is described in terms of both body schema and body image. How do you account for this and is there any other evidence to support these ideas?

You should now watch the video on proprioception, *Sense of proprioception: the man who lost his body*. Further details are given in the Block 5 *Study File*.

Proprioception

4

Having read Reader Chapter 19 and watched the video, you should already have some insight into this sense that contributes so strongly to our body schema. It is the sense that enables us to maintain posture and balance and to execute movements of the whole or any part of our bodies. **Kinesthesis** is a term sometimes used to denote *dynamic* proprioception or awareness of how the position of the body and of its parts are moving through space relative to one another. This would be distinct from a *static* position sense. However, the term proprioception quite properly covers both static and dynamic aspects of this sense.

4.1 Properties of proprioceptors

The mechanoreceptors that contribute most strongly to this sense are found in joints, muscles and tendons.

There are four kinds of joint receptors (Figure 4.1). Each receptor differs from the others in both its function and its anatomy, so that by the continuous integration of information about speed of movement, deviation from the preferred orientation, and the fact that the joint has reached the limit of its potential, we can both initiate and control our movements appropriately. Sensation provided by the four joint receptors works in tandem with input from the **Golgi tendon organs** and the **muscle spindles**.

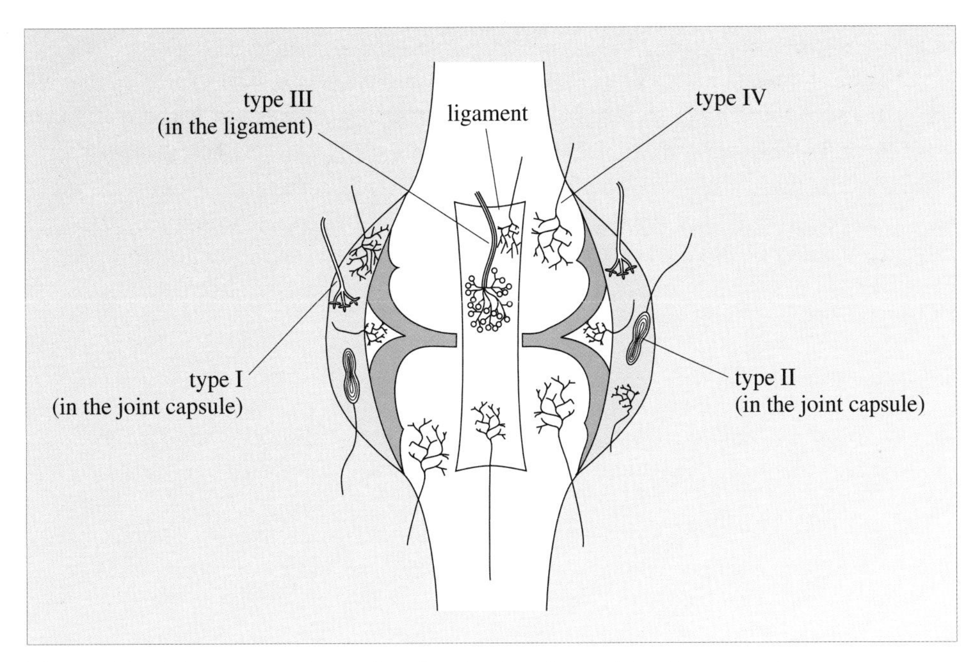

Figure 4.1 Diagram of a joint showing the locations of the four receptor types.

Whilst the joint receptors signal the movement of the joint, Golgi tendon organs detect the force of muscle contraction because they lie encapsulated between the muscle and tendon (Figure 4.2 overleaf). By contrast, the muscle spindles (Figure 4.3 overleaf) lie within the muscles and they can signal the stretching of a muscle.

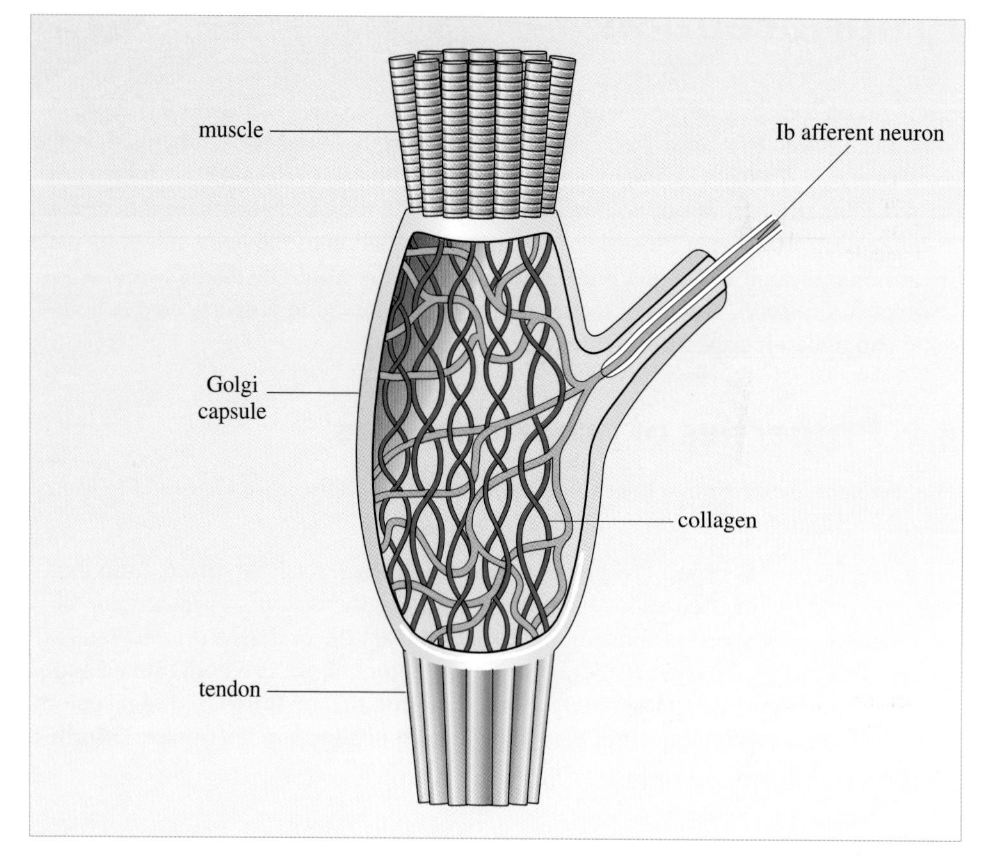

Figure 4.2 Diagram of Golgi tendon organ. The Golgi tendon organ lies within a capsule between a small group of muscle fibres and the tendon to which they attach. The Ib afferent neuron branches into free nerve endings within the capsule and these intertwine with the collagen fibrils connecting muscle to tendon. As the force of a muscle contraction increases, more groups of muscle fibres are recruited and so the number of Golgi tendon organs that are active increases.

The muscle spindle contains a type of muscle called intrafusal from the Latin *fusus* meaning spindle, so-called because these organs look like the spindles used by weavers. The intrafusal muscle is innervated by the free nerve endings of two types of afferent neurons. They can be distinguished because of a size difference. The large axons have diameters within the range 12–20 μm and are known as Ia fibres, the small afferents are termed group II fibres and have diameters from about 6–12 μm*. Both types are myelinated.

* At around the same period that these fibres were being investigated similar investigations were being made of the peripheral nerves close to the skin surface. Here a different nomenclature was used. The big fibres (12–20 μm) were called Aα and those of 6–12 μm were Aβ. These two systems of naming fibres have stuck!

❍ How will the function of the fibres differ?

● They will conduct action potentials at different velocities. The large fibres conduct more rapidly than the small fibres (Block 2, Section 2.1.1, Table 2.2).

The fine endings of each receptor type entwine around different areas of muscle as shown in Figure 4.3b. As long ago as the early 1930s recordings from individual axons had shown that the large Ia fibres detect the initial stretching of the surrounding muscles, increasing the rate at which they fire action potentials as the stretch develops but falling back to a steady rate of firing once the muscle maintains a particular length. By contrast the group II fibres respond more slowly as the stretch develops and they maintain their firing rate throughout the stretch (Figure 4.4). (There are also static and dynamic fibres involved in muscle innervation; see Reader Chapter 20, *Proprioception.*)

▶ **Figure 4.4** The responses given by muscle spindle afferent fibres when muscles are being passively stretched. (a) A diagram of the recording arrangement; (b) the traces obtained for muscle tension and firing patterns of the spindle afferent fibres.

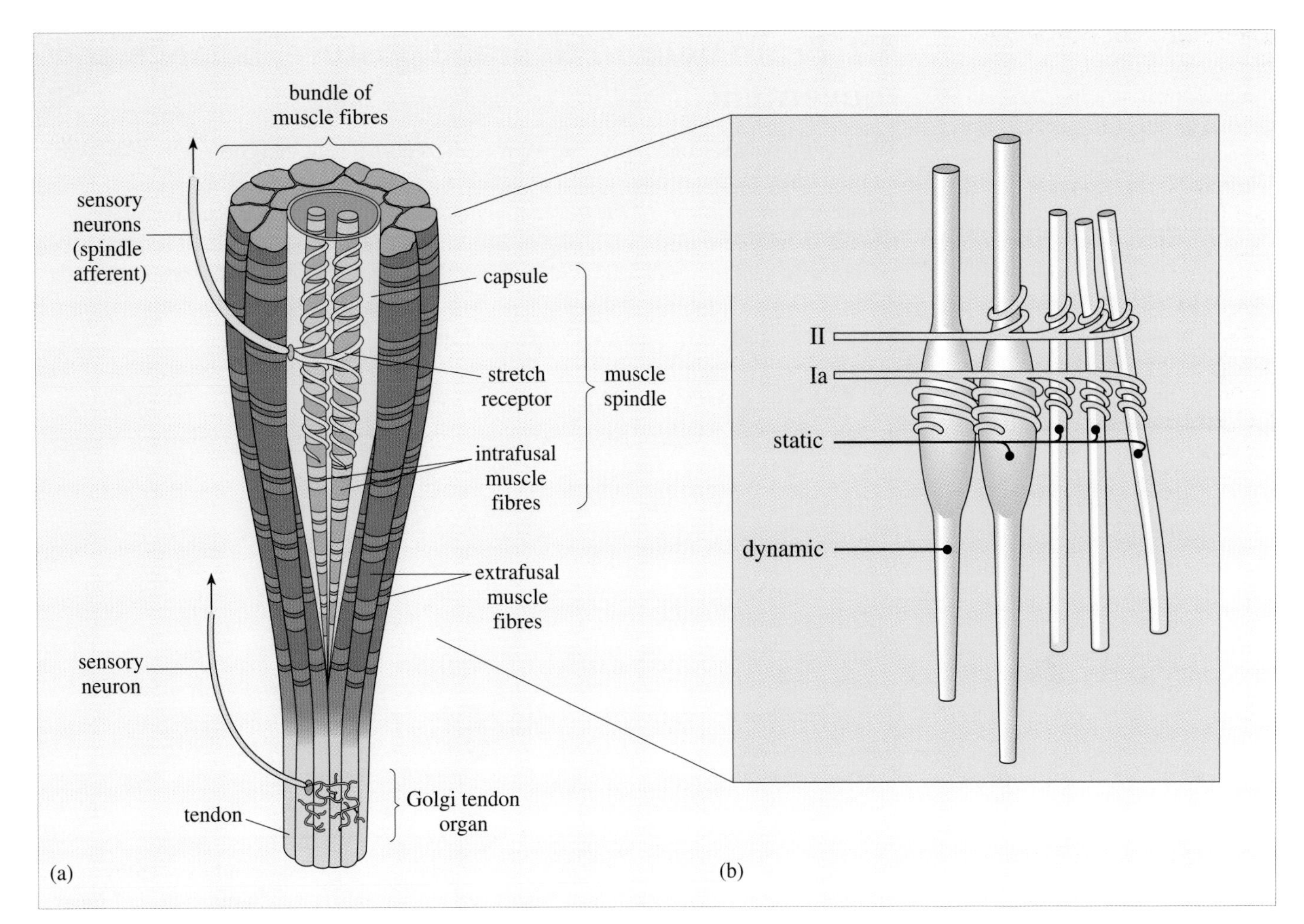

▲ **Figure 4.3** (a) Diagram of a muscle spindle showing how it lies encapsulated in skeletal muscle within a group of extrafusal muscle fibres. (b) Diagram showing the Ia afferent neurons with their free nerve endings wrapped around the central portions of the intrafusal muscle fibres and the II afferent fibres with their free nerve endings wrapped around the extremities of the intrafusal fibres. The static and dynamic fibres shown in (b) are motor fibres that are discussed in Chapter 20 of the Reader.

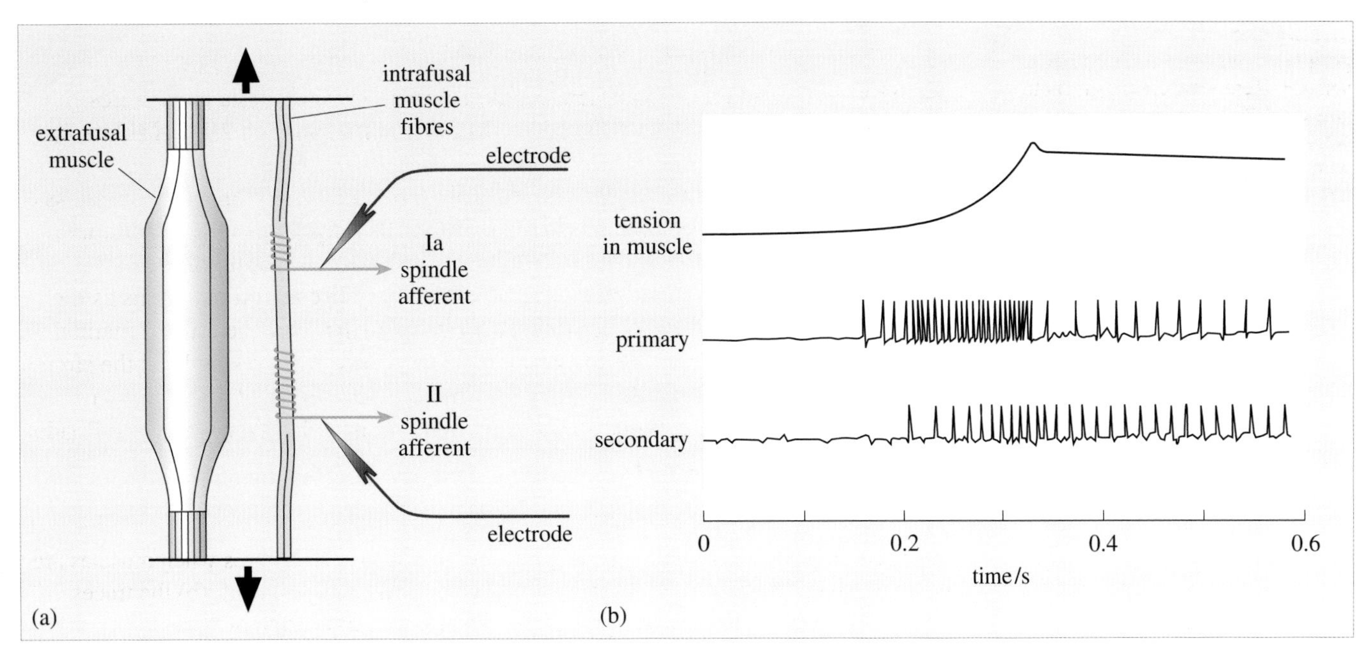

4.2 From neuron to brain: posture, balance and movement

The receptors described above are all modified dorsal root ganglion cells and feed into the same pathways described in Block 2 and in Chapter 18 of the Reader, *Somatosensory coding*. At the cortical level there is parallel processing with the muscle spindle afferents projecting to Brodmann's area 3a, whilst joint receptor afferents project to area 2. The video should have convinced you that when all is going well for us the visual contribution to balance and movement is secondary but when things go wrong it assumes a major role.

❍ Which other sense is important for the control of posture and movement?

● The vestibular system (Block 3, Section 6; also Chapters 7 and 28 of the Reader).

Activity 4.1 A balancing act

The importance of the integration of vision with our perception of body position when performing dynamic activity involving balance can easily be demonstrated. First stand still, then close your eyes. Easy? Make the task a bit more difficult by standing on one leg and closing your eyes. You should try this now but be aware that some people lose their balance very rapidly (I do!) so make sure you have some props (table, chairs, the wall) nearby to assist you in regaining your balance and avoiding a fall. Try to notice what goes wrong and how you attempt to correct it before reading on.

Even when standing on both legs we are not very stable because our centre of mass is very high, being somewhere around the hips. So a slight movement can cause the centre of mass to move beyond the base provided by your feet and you start to topple. However, a range of receptors communicate this unsatisfactory state of affairs to your brain. For example, somatosensory receptors in the two feet will be reporting a difference in pressure as you start to topple.

❍ How will the vestibular system contribute?

● The vestibular apparatus detects orientation and movement of the head (Reader Chapter 7).

It is clearly necessary for us to be able to distinguish the difference between head movements caused by the movement of the neck when you tilt your head to one side, and the same movement caused by the leaning of the whole body when you might be about to lose your balance (as in the activity above).

❍ Suggest how this distinction might be made.

● There are several possibilities. Given the particular activity above, your first suggestion might be that you hadn't issued any motor commands to tilt the neck so a moving head means the body is losing balance. Joint receptors in the neck and muscle spindles in neck muscles contribute information that tells us whether our neck and head are moving *relative* to the rest of the body or *together with* the whole body. When standing on one leg the difference in the pressure on the soles of the two feet will be huge, but there will also be detectable differences in pressure across the foot that remains on the ground and these should help to guide the muscular movements that must be made to keep you upright.

One thing is clear from the activity. Without vision it is far more difficult to keep your balance when standing on one foot. You probably felt yourself swaying and the foot and leg muscles working rather hard to keep you upright. Why should lack of visual input make such a difference?

The answer to this question is not really known, but balancing on one leg is a tricky operation and the senses work synergistically to achieve this. Removing one source of guidance cues (visual in this case) impoverishes our ability to act effectively. Notwithstanding this, balancing on one leg with your eyes closed is achievable. As you saw in the video, the reverse is also possible. Vision can compensate for the loss of proprioception – but only after considerable learning and effort.

You might have experienced temporary loss of proprioceptive input yourself. Have you ever experienced a limb that becomes numb ('goes to sleep')? This can happen, for example, when prolonged steady pressure cuts off the blood supply thereby impairing the functioning of nervous tissue. On release of the pressure the limb quickly returns to normal. But if you try to stand on a numb leg, it collapses because without the somatosensory feedback the brain does not know where it is and cannot control it. Episodes such as these give those of us with fully functioning somatosensory systems a tiny insight into the frightening world of life without proprioception. To find out more about this world, read Chapter 20 of the Reader, *Proprioception* by Ian Lyon now.

4.3 Summary of Sections 3 and 4

The somatic senses collect information from within the body and from our surroundings that makes possible our effortless interaction with our environment. We are consciously aware of only a tiny amount of this information but we do seek it actively in some situations such as when we explore objects with our hands. We integrate and use the information from many sources (i.e. receptors) to orient ourselves, position ourselves and to achieve a conscious awareness of the physical space around us and our place within it. The essentiality of these senses are demonstrated when they are lost through destruction of nervous tissue (neuropathy).

As with other senses, loss and illusion give insight into the complex processing and interactions between senses that give rise to perceptual experiences. They have also given rise to two terms that are conceptually useful to an understanding of the body senses: body schema, which means knowing where all the bits of your body are at any one moment and body image which conveys the sense of awareness of our physical attributes.

Question 4.1

(a) In both Chapters 19 and 20 of the Reader you met Jacques Paillard's patient who had damage to the left parietal lobe. This resulted in no conscious awareness of any somatosensory input from the right arm. Draw a diagram to show the subcortical areas to which the sensory afferents from the right arm project.

(b) How would you account for the fact that the patient could move her left arm to indicate the exact place where the experimenter touched her unseen right arm when the experimenter said 'here'?

5 Pain

Pain is, by definition, unpleasant. In Block 1 you were alerted to the fact that this sets pain apart from other sensations. The stimulus that gives rise to the sensation of pain is, most often, tissue damage. Tissue damage is something to be avoided, and pain can be useful in alerting us to the damage so that we can take immediate action to ameliorate its effects and perhaps also learn to avoid similar episodes in the future. Once we have recognized that damage has been done, it is necessary to contain the damage and to start the healing processes as soon as possible. When the pain has done its job of communicating to us that we have a problem, we would like it to stop. But it often continues. In the short term that may be a good thing because it may *force* us to take things easy thereby allowing the body's finite resources to be pumped into repair and restitution. But when pain persists long after healing is apparently complete, it is recognized as a clinical syndrome and a problem in its own right.

Thus it is that the focus of research into pain is somewhat different to that undertaken into the other senses. What we are most interested in, these days, is how we can control or relieve pain. One way to approach these challenges is to find out more about the neurophysiology and neuropharmacological aspects of pain. This is the focus of Reader Chapter 21, *The neurophysiological basis of pain*, which you will be asked to read in Section 5.4.

The history of the study of pain shows that the control of pain has not always been the guiding principle for research and a little investigation shows us why this is the case. The word pain derives from the Latin *poena* meaning punishment. That pain was a punishment for past wrongdoings was a belief held by many peoples from diverse cultural backgrounds. The consequence of this belief was that you suffered pain, i.e. you put up with it in an uncomplaining fashion.

There are a number of reasons for the belief that pain is inflicted for past misdemeanours. One reason is that it has always been a natural human attribute to look for causes and as pain is not always correlated with physical damage some other cause needed to be found. Logically an unpleasant sensation must have an unpleasant cause and so the pain was believed to be the fault of the sufferer.

This dissociation between the stimulus and the sensation can be a **double dissociation**, because in some cases where tissue is damaged, no pain is reported (Figure 5.1 overleaf). This is particularly the case when there is a high level of arousal (Block 1, Section 2.3). So on the one hand one might see soldiers, terribly wounded in a glorious struggle, dying without complaining of pain, and on the other hand there might be an individual with no apparent injury, in terrible pain, and who subsequently dies – ingloriously. In the latter case there might not have been a true dissociation, it could have been that there was internal damage and disease. There are certainly a great many cases now where 'pain without injury' would be deemed an incorrect diagnosis because the cause of the pain has been established as **neuropathic**. Neuropathic pain is pain attributed to damage or malfunction of the nervous tissue itself. But this does not completely account for the problems of the double dissociation of tissue damage and pain. There are still cases where pain is experienced for which no physical correlates can be identified and also there remain a few individuals who, in spite of tissue damage, cannot feel pain even though they have an apparently normal nervous system (Block 1, Section 4.4).

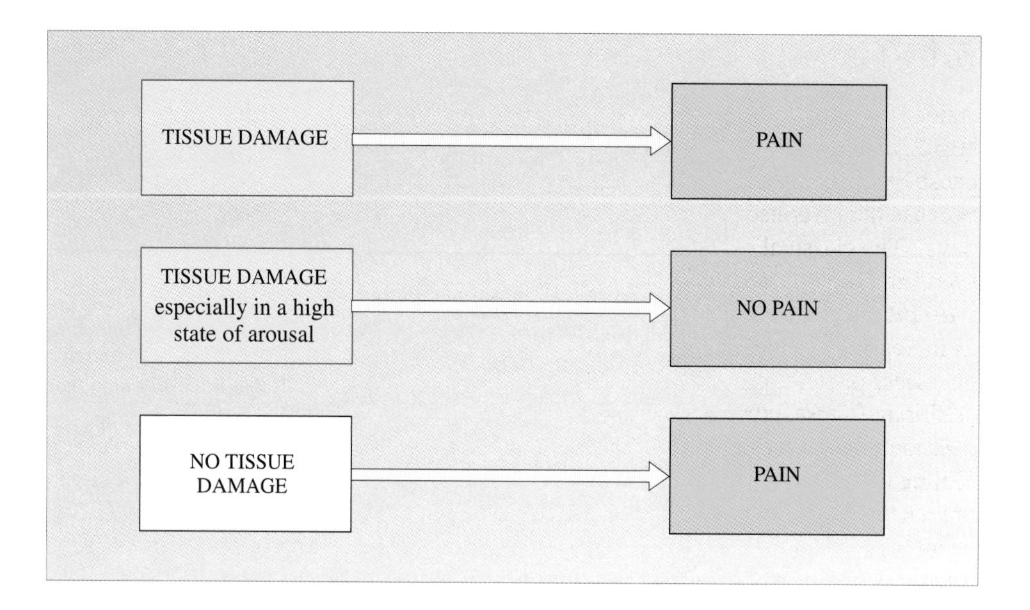

Figure 5.1 The double dissociation of tissue damage and pain.

The legacy of this misunderstanding of pain can be seen to have lingered on until quite recently. This was probably not helped by the fact that outside of the battlefield, the groups that regularly experienced pain were women menstruating (period pains) and women in labour (childbirth) whilst most medical doctors were men and most scientific research was done by men.

A final word on the history of pain control (or rather, lack of it) can be made from a slightly different angle, the fear of addiction. Many chemical substances that are found in plants have **analgesic** (pain reducing) properties and have been known and used for centuries for the relief of pain. However, some of the most powerful of these substances are used recreationally because they, variously, induce hallucinatory or euphoric experiences. They also can become addictive when used in this way. So modern medical practice, having become divorced from the wisdom of the ancient herbalists, assumed that these substances given for pain relief would also induce addiction in the patient and so pain relief was withheld. However since the 1970s research evidence has accumulated showing that substances such as the opioids (derived originally from the opium poppy) do not become addictive for patients using them to control or relieve pain.

This view was championed back in the 1960s by the British physician and researcher Patrick Wall. He, together with Ronald Melzack, developed a very influential theory encompassing the mechanisms by which the body could damp down the sensation of pain naturally and how we might, by understanding this, more effectively reduce pain by artificial means. This is known as the **gate-control theory of pain**. It postulated a 'gate' at the level of the spinal cord within the dorsal horn that could regulate the effects of afferent axons from nociceptors that were signalling tissue damage. If the gate were closed there would be no pain. But if the gate was open the afferents could excite the secondary neurons that projected up the spinal column to the brain and there would be a perception of pain. This theory was very important in driving forward pain research. The details of the mechanisms proposed by Melzack and Wall have not been substantiated but the concept remains valid and in Chapter 21 of the Reader you will read how current understanding of the neuroanatomy and neurophysiology of the spinal cord fits with the gate-control theory of pain.

5.1 Tissue damage

Aside from cuts and bruises, tissue damage can be caused by bacterial and viral attack, chemicals and heat. The injured tissues release a variety of substances that cause local changes collectively known as **acute inflammatory response** (acute because the response usually lasts only a matter of hours and at most for a few days). The classical components of inflammation are pain, heat, redness and swelling. The inflamed area shows these signs because the local blood vessels dilate (*vasodilation*) thus increasing blood flow into the area, causing it to look red as well as increasing the local temperature. The permeability of the capillaries increases and an excess of fluid, proteins and cells leak into the area that becomes swollen by the addition of these extra materials. An early effect of inflammation is a clotting reaction which blocks surrounding tissue spaces and lymphatic vessels effectively sealing off the area. (Later this may manifest itself as bruising.) Sealing off the area prevents infection spreading to other tissues. Interestingly the speed with which this occurs is in proportion to the severity of the tissue damage, and so staphylococcal infections, such as *Staphylococcus aureus*, which release highly toxic products causing cell death and abcess formation, will be responded to more rapidly than the less virulent streptococcal infections, such as tonsillitis and pneumonia. Hence streptococcal infections are, in general, more likely to spread around the body and even to cause deaths.

Cells of the immune system also contribute to the inflammatory response and the most significant of these are white blood cells (leukocytes) called **mast cells**. Mast cells are manufactured in bone marrow, released into the blood and a few hours later they will be in the tissues. Their lifespan is only 4–5 days but will be shorter if they are triggered into *degranulating*, i.e. releasing granules of irritant chemicals, such as histamine which triggers vasodilation.

Vasodilation also occurs in response to the presence of two neuropeptides, substance P and calcitonin gene-related peptide, which are released from the tips of damaged nociceptors (Figure 5.2 overleaf). The two peptides also stimulate the degranulation of mast cells which is one of the ways that an involvement of the immune system can be triggered in the absence of bacterial infection.

Damage to cells (by whatever means) prompts the release of other chemicals many of which affect the firing of nociceptors. These include prostaglandins, kinins, serotonin (5HT), ATP, acetylcholine (ACh) as well as ions such as potassium (K^+) (Box 5.1 overleaf).

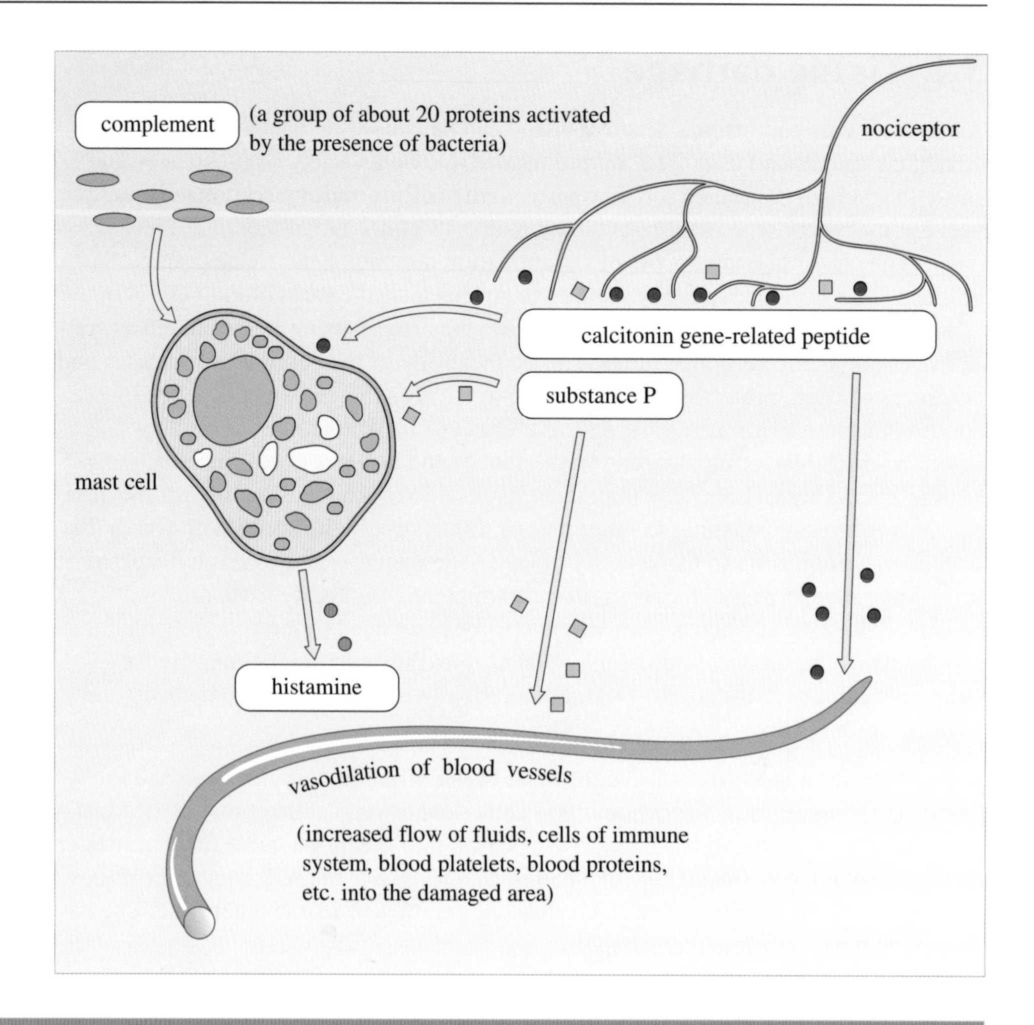

Figure 5.2 Factors that can trigger inflammation.

Box 5.1 The chemistry of inflammation

A part of the immune response, inflammation occurs when soft tissues are injured by, for example, trauma, bacteria, viruses, corrosive chemicals, and excessive heat or cold. The spread of the acute inflammatory response following injury suggests that chemical substances are released from injured tissues, spreading outwards into uninjured areas. Indeed, harmful stimuli such as these do trigger the release of certain chemical mediators of which two of the main families are the *kinins* and the *prostaglandins*.

The kinins

The kinins are small peptides with wide-ranging properties that include the ability to cause vasodilation, increase vascular permeability, bring about the contraction of smooth muscle and stimulate the arachidonic acid cascade (see later). They also cause pain. The best known kinins are bradykinin and kallidin (Figure 5.3). In addition to the properties listed above, there is recent evidence that bradykinin also plays an important role as a mediator of the beneficial cardiovascular effects of the important angiotensin-converting enzyme (ACE) inhibitors that help to lower blood pressure. Bradykinin is also implicated in allergies and asthma and is a potent stimulator of nociceptors. The process that forms bradykinin in the body is summarized in Figure 5.4.

(a) H—Arg-Pro-Pro-Gly-Phe-Ser-Pro-Phe-Arg—OH

(b) H—Lys-Arg-Pro-Pro-Gly-Phe-Ser-Pro-Phe-Arg—OH

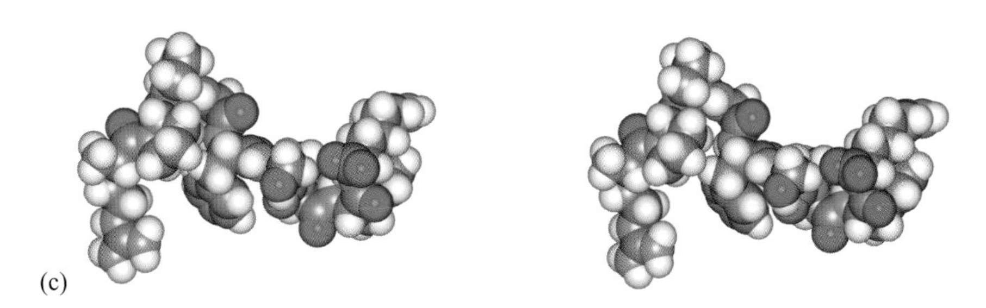

Figure 5.3 The structures of (a) bradykinin (a nonapeptide) and (b) kallidin (a decapeptide also known as lysyl-bradykinin because of its close relationship with bradykinin). (c) Stereoscopic space-filling model of bradykinin. (You should use the stereoscopic viewer for this figure.)

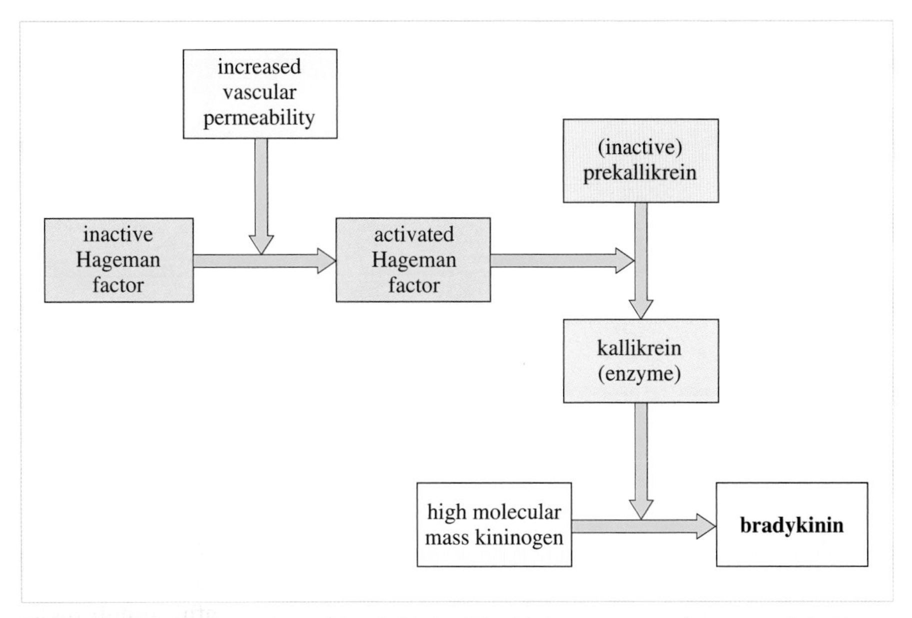

Figure 5.4 The formation of bradykinin. The kininogens are plasma α-globulins. There are two forms: high molecular mass kininogen (HK) and low molecular mass kininogen (LK). HK circulates in the plasma bound to prekallikrein, the precursor to kallikrein, an enzyme that releases bradykinin from HK. Prekallikrein is converted to kallikrein in a variety of ways. In the context of inflammation, the main activator is Hageman factor (factor XII of the blood clotting sequence). Hageman factor is normally inactive, but is activated by injury or any other cause of inflammation. The increased vascular permeability that occurs during inflammation allows the various components to leak out of the blood vessels. This activates the Hageman factor, which in turn converts prekallikrein to kallikrein; this then leads to the generation of bradykinin from HK by cleaving the Lys-Arg and Arg-Ser bonds that hold the nonapeptide chain within the HK molecule (the two Arg residues being the ends of the bradykinin molecule).

The prostaglandins

As mentioned earlier, bradykinin and other kinins stimulate the arachidonic acid cascade, giving rise to various prostaglandins. These are extremely potent substances that have many properties: some prostaglandins help protect the body (for example, they help the stomach resist ulcer formation); others contribute to the inflammation process and the accompanying pain sensation (they make the nociceptors more sensitive, thereby increasing the pain caused by the action of the bradykinin).

The prostaglandin portion of what is often called the arachidonic acid cascade is shown in Figure 5.5. (You do not need to remember these structures, they are shown simply to illustrate how minor changes can have major effects on their biological properties.) As you may recall from Block 2, arachidonic acid is stored within the cell in the form of cell membrane phospholipids. It is released as required by the action of an enzyme, phospholipase A_2.

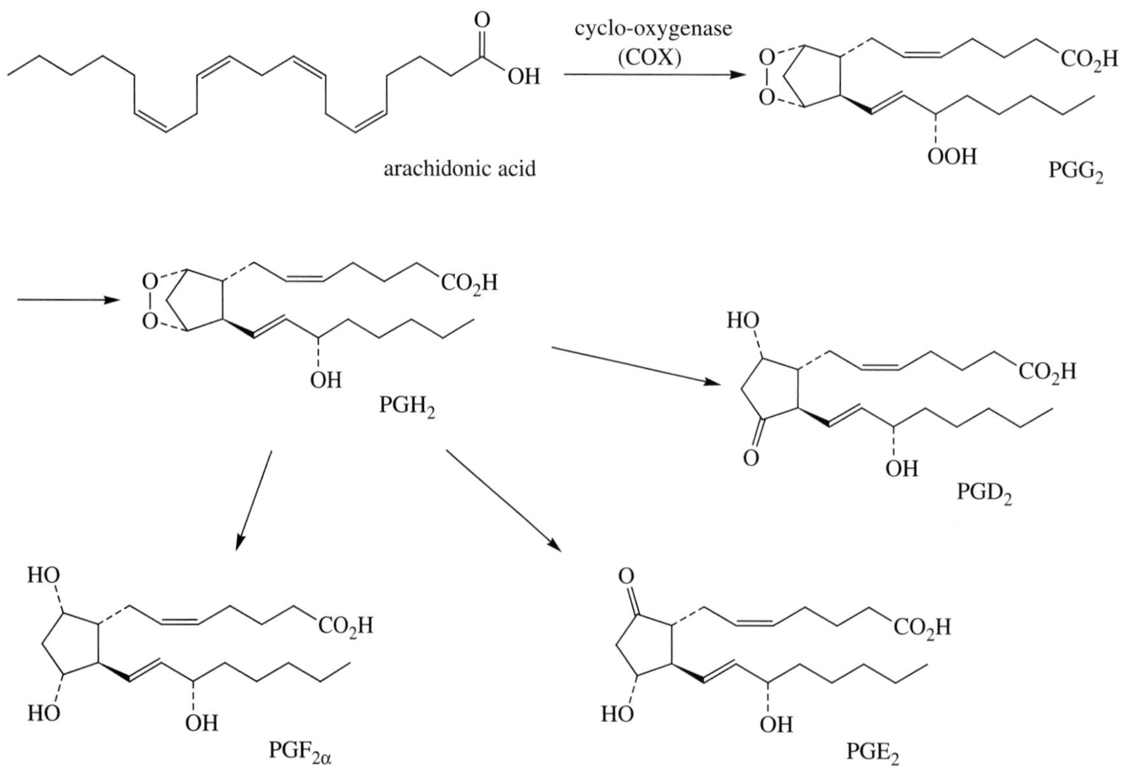

Figure 5.5 The production of prostaglandins from arachidonic acid. The enzyme cyclo-oxygenase (abbreviated COX) converts arachidonic acid to an intermediate known as prostaglandin PGG_2 (the nomenclature need not concern us, but is historically-based). This is converted by another enzyme to prostaglandin PGH_2. This in turn is transformed by three different enzymes into three possible products, each with different properties, PGD_2, PGE_2 and $PGF_{2\alpha}$. These are just a few of the very many metabolites formed from arachidonic acid.

The most common disease involving inflammation pain is arthritis. Arthritis is one of the world's most painful diseases, more common than heart disease, cancer or diabetes. To relieve the crippling pain of arthritis, millions of people around the world rely on non-steroidal anti-inflammatory drugs (NSAIDs). The oldest and best known of these is aspirin. Aspirin used to be the standard against which other NSAIDs were compared but now ibuprofen, which is safer, has taken on that role. NSAIDs such as aspirin, ibuprofen and other newer drugs like naproxen act by inhibiting the COX enzyme. Unfortunately, this prevents the formation of beneficial prostaglandins as well as acting as a pain-reliever. Consequently, each year one in every 1200 people who take these medications for longer than two months die as a result of gastrointestinal and renal side-effects.

In 1991 it was discovered that there are two cyclo-oxygenases, called COX-1 and COX-2: COX-1 is present all the time and is largely responsible for producing the beneficial prostaglandins; COX-2 is formed in response to an inflammation stimulus and is responsible for the painful effects. This raised the prospect that it might be possible to target just COX-2, and indeed two drugs that are selective COX-2 inhibitors (rofecoxib and celecoxib) became available in the late 1990s.

5.2 Properties of nociceptors

Nociceptors are dorsal root ganglia cells whose peripheral receptive elements are finely branching. These free nerve endings are of at least three types (Table 5.1). Additionally there are visceral nociceptors known as silent nociceptors because they are rarely active. However, they are important because they may be responsible for some of the chronic, nagging pains that are experienced.

Table 5.1 Types of nociceptors.

Name	Stimuli to which they respond	Fibre type
polymodal nociceptor	chemical thermal mechanical	C
thermoreceptor	temperatures $<5\,°C$ $>45\,°C$	Aδ
mechanoreceptor	intense pressure, e.g. pinching	Aδ

The mechanisms by which the stimuli are transduced are obviously varied, particularly in the polymodal receptors. This is an area of intensive research activity. Any kind of actual tissue damage changes the damaged cells which include nociceptors. So nociceptors change over time and the transduction processes may change too.

We have already discussed the fact that broken or damaged cells will release chemicals that affect other cells in their neighbourhood in a number of ways. The released substances may have one (or more) of three effects. First, they may have direct effects upon nociceptors firing rates, and second, they may sensitize the nociceptors so that they become more readily active when they encounter noxious (or even previously innocuous) stimuli (Figure 5.6 and Table 5.2). This latter effect is the basis for two phenomena that you are very likely to have experienced yourself. An injury such as a burnt or bruised hand is very sensitive, the least knock will cause intense pain. This is described as **hyperalgesia**. But more surprising is the pain elicited by an innocuous stimulus such as running warm water over a burned skin area (see Chapter 21). This is known as **allodynia**. The situation is often made worse by inflammation, which is the third effect brought about by those substances released from damaged cells (see Figure 5.6 and Table 5.2) and has already been discussed.

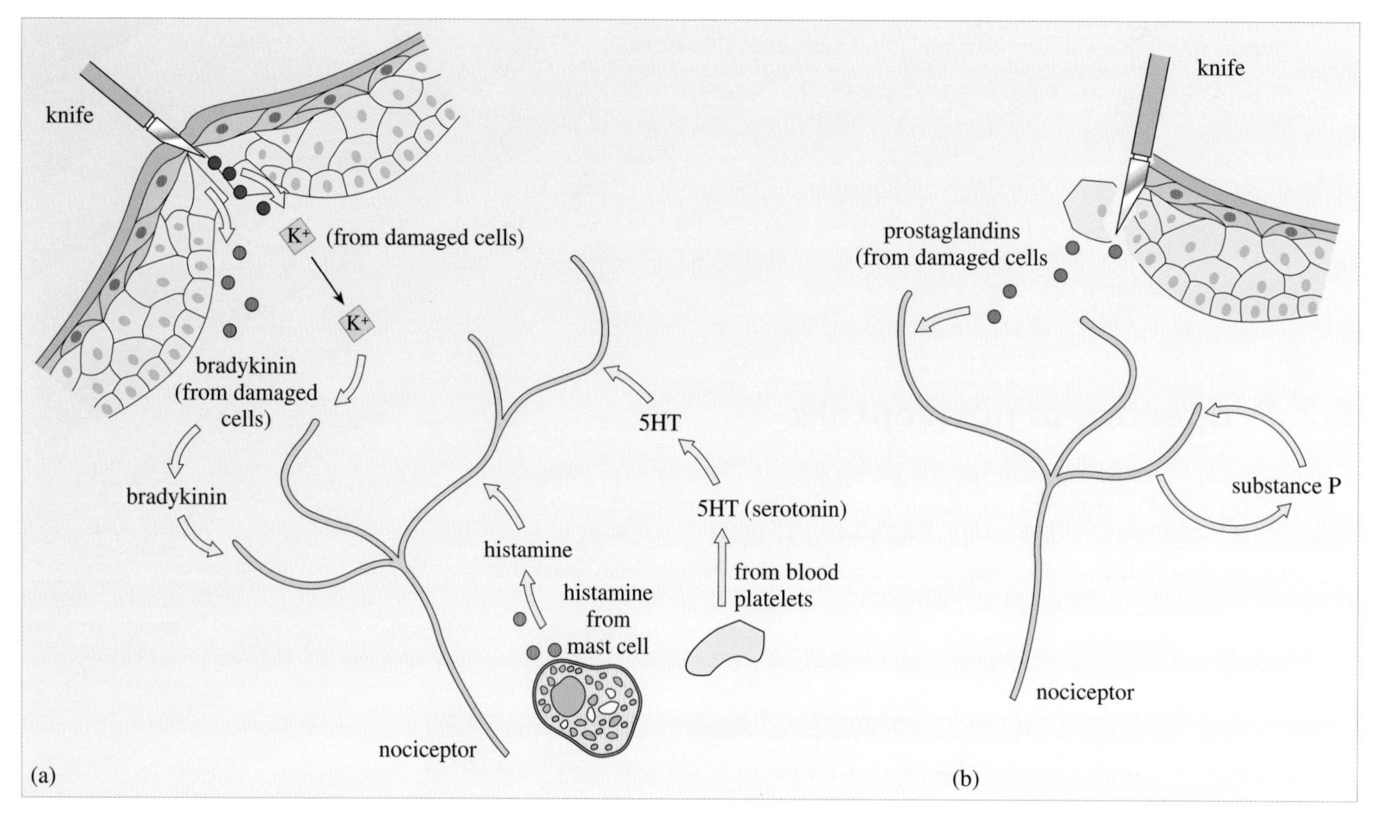

Figure 5.6 (a) Direct effects on nociceptors. Release of these substances causes nociceptors to fire. (b) Indirect effects on nociceptors (these lower the firing threshold of the nociceptor, i.e. they sensitize).

Table 5.2 The effect on nociceptors of some chemicals released by damaged cells.

Chemicals that activate	Chemicals that sensitize
potassium	prostaglandins
histamine	substance P
bradykinin	acetylcholine (ACh)
serotonin (5HT)	adenosine triphosphate (ATP)

Box 5.1 Responses to heat: the phenomena of allodynia and hyperalgesia

Let us now consider the responses from peripheral receptors in the skin as we experience first warmth and then pain from standing too close to a fire. Figure 5.7 shows the response of a polymodal nociceptor to heat stimuli. Thermal nociceptors also respond to temperatures in this range.

❍ Compare the response of the nociceptor shown in Figure 5.7 with the response of a 'warm' thermoreceptor shown in Figure 3.2. How do they differ?

● A 'warm' thermoreceptor stops firing at temperatures greater than about 45 °C. The nociceptor has a heat threshold somewhere around 43 °C and continues firing at temperatures up to 58 °C.

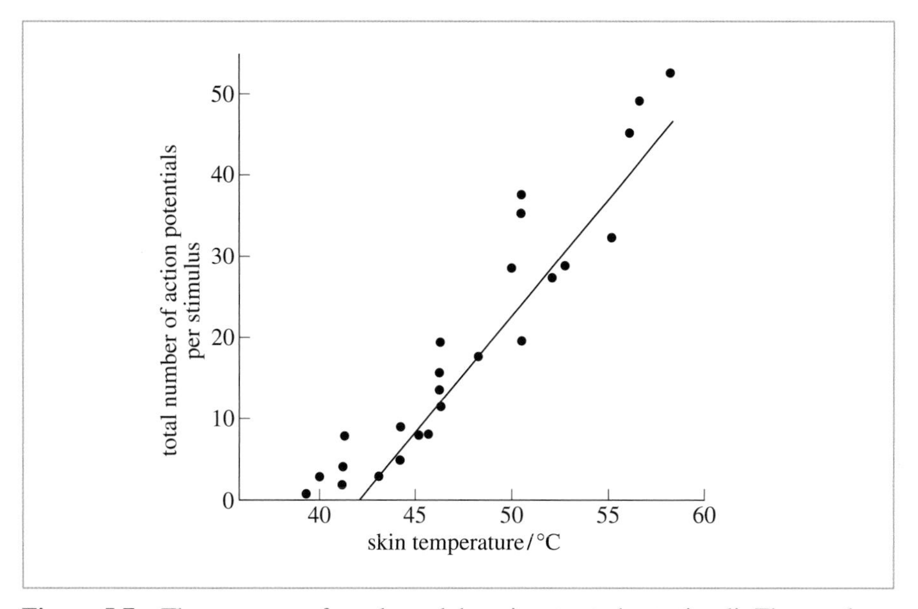

Figure 5.7 The response of a polymodal nociceptor to heat stimuli. The graph shows temperature plotted against the number of action potentials evoked by various stimulus temperatures.

The perception attached to the firing of these receptors also differs. As the temperature rises the increased firing of the thermoreceptor is perceived as increased warmth whereas the increased rate of firing of the nociceptor is experienced as more intense pain. Most of us know to our cost that even if we remove ourselves very rapidly from a painful heat source actual tissue damage will already have occurred. We will have burnt skin. Figure 5.8 shows how this initial exposure to the noxious heat stimulus affects the subsequent responses generated by the same stimulus. The nociceptors have been sensitized and both hyperalgesia and allodynia are exhibited. So if we withdraw from the fire, then move close again we will experience pain at a temperature that previously just felt pleasantly warm (allodynia). And the temperature which first gave rise to the experience of pain will now be experienced as more painful than before (hyperalgesia).

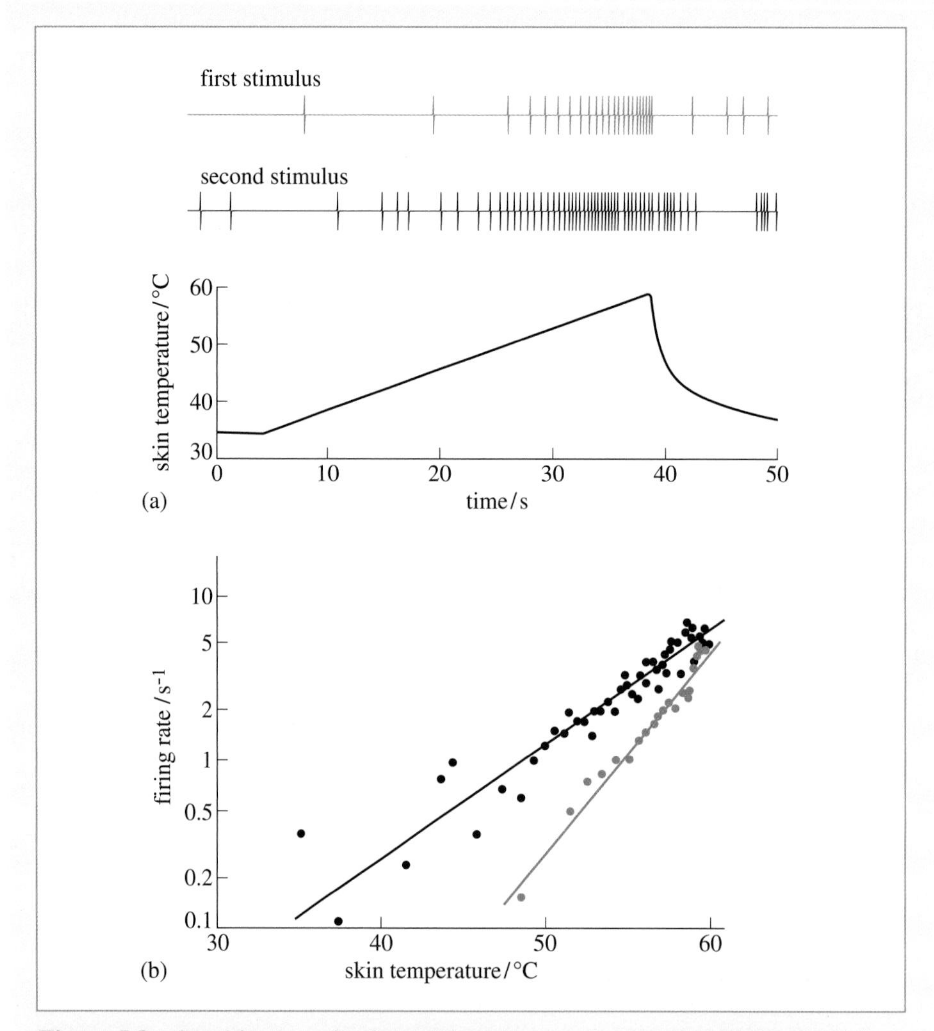

Figure 5.8 Sensitization of polymodal nociceptors by repeated application of noxious heat stimulation. (a) Nociceptor responses to the heat stimulus. The upper trace (blue) shows the action potentials generated by the first application of the stimulus and the lower trace (black) shows the response to a second application made 4 minutes after the original stimulus. The graph below shows the change in skin temperature during the application of the radiant heat stimulus. The skin temperature is increased progressively from an initial level of 34 °C to just under 60 °C in 35 seconds. (b) Graph of the response (the action potential frequency) plotted against skin temperature. At any given temperature, the magnitude of the response to the second stimulus (black dots) is greater than the response to the first stimulus (blue dots).

A glance at Table 5.1 shows that painful thermal stimuli (above 45 °C and below 5 °C) are not only transduced by polymodal nociceptors but also by dedicated thermo-nociceptors having Aδ fibres. These small diameter fibres have a thin layer of myelin, whereas the C fibres of the polymodal nociceptors (whose firing pattern is shown in Figure 5.8) are even thinner and unmyelinated.

❍ How do axonal (fibre) diameters affect speed of conduction?

● The smaller the diameter, the slower the conduction velocity (Block 2, Table 2.2).

❍ In what way does myelin affect the conduction velocity?

● Myelinated axons conduct action potentials more speedily than unmyelinated axons (Block 2, Section 2.1.2).

This means that for thermal and for mechanical stimuli (but not for chemical stimuli – see Table 5.1) there are two types of receptor able to signal noxious stimuli and one conveys the information more rapidly than the other. Thresholds are also different. The polymodal nociceptors only respond to the more intense stimuli. This suggests that there may be two types of pain perception for any one noxious stimulus.

Personal experience confirms two types of pain are perceived after an acute insult, such as a cut or a burn. The *first pain* (or fast pain) is often described as sharp, bright or stinging pain whereas the *second pain* (or slow pain) is more intense, a dull ache that is more diffuse, harder to localize. Direct stimulation of the fibres gives the same results. Stimulation of Aδ fibres elicits reports of sharp, stinging pain whereas stimulation of C fibres is reported as an intense dull ache.

❍ What might be the function of two sets of receptors responding to noxious stimuli?

● The more rapidly transmitted signals in Aδ fibres warn of the onset of noxious events. The Aδ input triggers withdrawal responses. The more slowly conducted signals in C fibres signify that tissue damage has occurred (the chemical cocktail released by damaged tissues (Section 5.3) activates only C fibres via polymodal nociceptors). The sensations engendered by C fibre discharge encourages immobility and nurturing of the damaged area.

The relationship between stimulus and perception is not as straightforward as the above description would suggest. If you put a tourniquet (a very tight 'bandage') around a limb the blood supply is reduced to the point of anoxia and only C fibres can continue to function (because they have the lowest metabolic rates). Now the perception is of burning pain regardless of whether the stimulus is a pin-prick, ice or a pinch – all stimuli that are normally easily discriminated.

5.3 Transduction of noxious stimuli

Previous mention was made of this being an area of intensive research effort. Clearly the better we understand the processes the better the chance of designer drugs for pain relief. We are only at quite elementary levels of understanding at the moment. In many cases we use analgesics whose properties have been discovered serendipitously and we do not know how they work.

As an example, consider chilli peppers! They contain an ingredient *capsaicin* that can cause skin irritation when the peppers are handled and that excites the internal mucosa (skin) of the mouth and gut to a point that many find unbearable. (The rest of the population just enjoy the hot, spicy zing it adds to their food.) Capsaicin is also used in a cream as an analgesic particularly for arthritis. Various interesting suggestions have been made for its efficacy in the absence of any 'hard facts'. Gradually the story is emerging, though it is still incomplete at the time of writing (2002). There are ionotropic receptors called VR-1 in at least some polymodal nociceptors that it has been possible to clone and study. These ligand-gated ion channels allow the entry of the cations Ca^{2+} and Na^{+}.

❍ What significance attaches to the opening of cation channels by these receptors?

● Na^{+} and Ca^{2+} ions will enter the cell down electrochemical and concentration gradients and depolarize the nociceptor. If depolarization is sufficiently large the cell fires and we perceive pain or itchiness.

The VR-1 receptor site is on the inside of the cell so capsaicin has to pass through the neuronal membrane to activate the ion channel. The channel is also activated by heat, acid and anandamide (Figure 5.9).

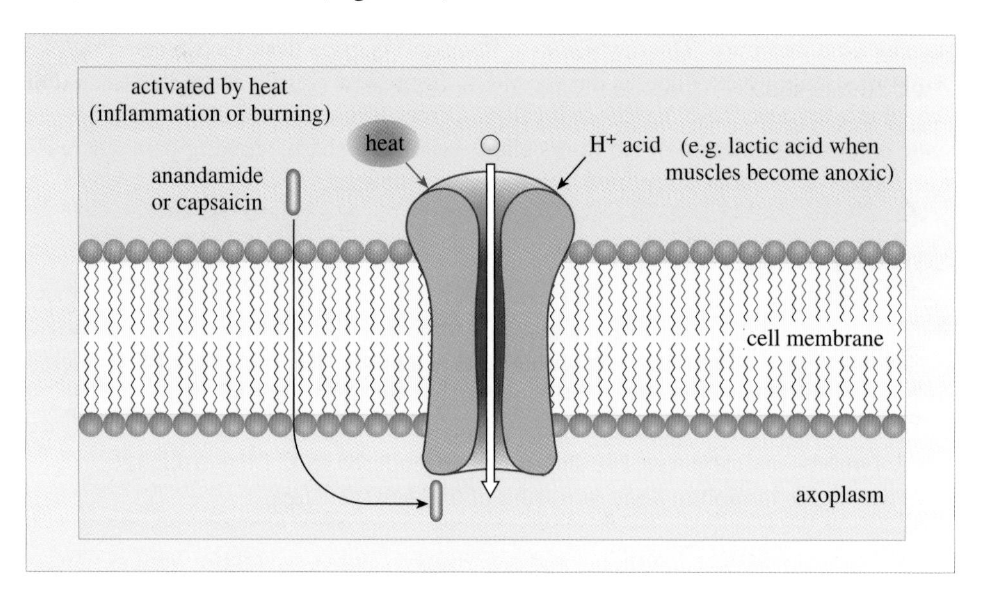

Figure 5.9 A schematic diagram of the VR-1 receptor.

Anandamide is an endogenous cannabinoid (i.e. a substance found in the body that has marijuana-like properties). The active ingredient of marijuana is delta-9-tetrahydrocannabinol (THC). Besides activating these peripheral nociceptors, which leads to the painful sensations reported (as for capsaicin), anandamide is a vasodilator, hence the initiation of an inflammatory response and subsequent hyperalgesia. However, repeated application of capsaicin peripherally gives pain relief and this seems to be because, in some way, it prevents neuromodulators such as substance P from being released peripherally (see Figure 5.2) and also from the nociceptor synaptic terminals in the dorsal horn of the spinal cord. So part of the biochemical story has emerged but there is still a long way to go for a complete understanding. You will read more about the afferent pathways and substance P in the two Reader chapters on pain.

There is another ionotropic receptor type that is currently under investigation which is found in some nociceptors and opens channels for Na^+ and Ca^{2+}. This receptor has been found to bind with ATP. ATP is released by rapidly growing tumours. ATP is also released by damaged cells, particularly in areas of **ischemia**. Ischemia is a condition where blood flow is substantially reduced. This happens for example when muscles around blood vessels go into spasms and it can be the cause of certain kinds of headache (particularly migraine) as well as angina and muscular cramps. Muscle spasms have a directly constricting effect on blood flow and possibly also affect mechano-nociceptors directly by pulling them. Of course once the nociceptors respond they may also release substance P from their terminals at the peripheral site of ischemia and the complex chemical cascade of sensitization, activation and inflammation will be triggered (see Figures 5.2 and 5.6).

It must be very clear by now that the interaction of the stimuli which give rise to activity in nociceptors is complex and we are really only at the beginning of an understanding of the transduction mechanisms. Whilst the immediate response to a noxious stimulus is often depolarization of the nociceptor, the nociceptor may itself be physically damaged and hence change its behaviour. There is also a longer-term effect on the neuron's immediate environment which will certainly be changed by the inflammatory response. The result of all this may be that there are changes in the genes expressed by the nociceptor. Which genes are affected and in what ways may depend on the precise nature of the original tissue damage and we could imagine that different stimuli might trigger a different set of biochemical changes in the cell. Watch this space!

5.4 Afferent pathways and the gate-control theory of pain

Nociceptors have their cell bodies in the dorsal root ganglia and from here axons enter the spinal cord via the dorsal roots. Once in the spinal cord they synapse with other neurons, which relay the signals to the brain via several ascending pathways. These pathways are described in Chapter 21 of the Reader, *The neurophysiological basis of pain* by Julian Millar which you should read now. You will probably find it helpful to read it straight through first and then to make your own annotated diagrams of the pathways that are implicated in the perception of pain so that you can answer the questions below.

After reading the chapter, you will appreciate that there is still a great deal to be discovered in the field of pain research. One thing that might have struck you is that the endogenous opiates (enkephalins) intervene at two levels in the nervous system (at a minimum).

❍ Where are these two areas?

● Enkephalins are active at the spinal cord level in the dorsal horn as well as in the midbrain to hindbrain PAG neurons in relation to pain phenomena.

These are not areas that are directly available to consciousness but experiences induced when the drugs are taken recreationally are consciously experienced.

❍ Where are the addictive and euphoric qualities of opiates believed to be acting?

● At the midbrain and hypothalamic opiate receptors (Chapter 21, Section 3.4).

So these are not the sites of conscious awareness either. The pathways that can be stimulated by opiates are varied: those that mediate analgesia are descending pathways whereas the euphoric qualities come from the stimulation of ascending dopaminergic pathways. Obviously addiction is a complex phenomenon and worthy of a course in its own right. (The Open University currently has an MSc course, *Issues in Brain and Behaviour*, which studies addiction as one of its options.)

Whilst addiction is not a problem for those in chronic pain, *tolerance* is. A patient is described as being tolerant to a drug if over time the patient finds that the dosage has to be increased to retain the benefits of the drug. This creates difficulties when the drug has other effects (as most do). The opiates tend to cause respiratory depression so prescribing appropriate dosages that will relieve pain without causing respiratory failure is very much more of an art than a precise scientific activity, because of the immense variations in the way that individuals respond to treatment. The next section will look at this in more detail.

5.5 The control of pain

Pain is a valuable diagnostic tool so that even after an accident or injury the questions: 'Does it hurt?' and 'What does it feel like?' tend to be asked before pain-killers are administered. We ask ourselves similar kinds of questions before deciding which type of analgesic to take: how much (one or two tablets?) and do I need to book an appointment with a doctor? Frequently, and especially in cases of acute tissue injury, the analgesic is dealing with the symptoms, not effecting a cure for the presenting problem. With an accurate diagnosis the presenting problem can usually be treated medically – from dressing a wound to operating for an appendicitis – but wounded and damaged tissues do not revert instantly to their previous state and C fibres show little adaptation in their firing rate. The continuing firing of the nociceptor C fibres is perceived as continuing pain. Thus any mechanism that prevents their firing should be accompanied by a reduction in pain.

Pain, other than superficial, cutaneous pain, can often be quite difficult to localize accurately. This is especially true in the case of visceral pain. As previously mentioned, the visceral nociceptors are mostly silent. We only become aware of visceral pain when nociceptors over a large area become active. The reason that this visceral pain is difficult to localize is because the visceral nociceptors synapse with the same second order neurons in the dorsal horn of the spinal cord as the peripheral nociceptors. It is therefore difficult for the brain to know which input led to the firing of the dorsal horn neurons. This 'wiring problem' may explain the phenomenon of **referred pain** (see Figure 5.10).

Figure 5.11 shows some common internal sites that, when damaged or inflamed, result in pain being experienced at ('referred to') a completely different location. The best known example is probably the patient with angina who reports pain in the left arm and shoulder although the site of injury is the heart. This apparent absurdity is not a haphazard association but one that reflects the developmental origins of the organs concerned and their innervation. Because associations are made in a consistent way based on these anatomical principles, referred pain can provide the clinician with useful diagnostic information. Referral is always in one direction. We do not normally 'hear from' our internal organs so when the 'silent' visceral nociceptors are active we interpret the activity as coming from the more familiar peripheral locations and not vice versa. Active visceral nociceptors can be

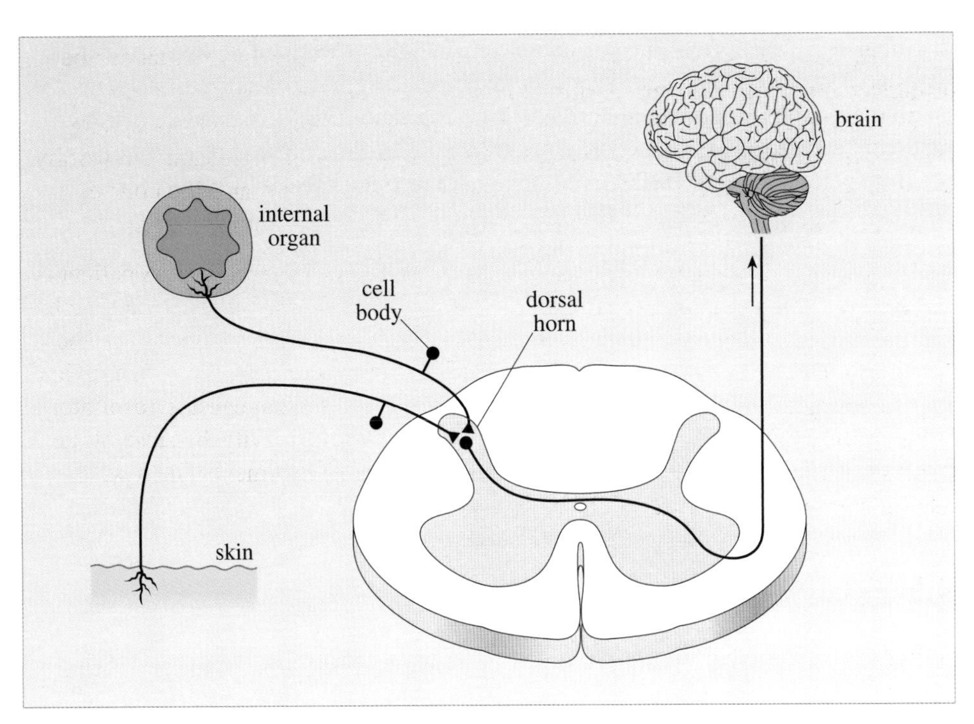

Figure 5.10 A suggested neural basis for the phenomenon of referred pain. Some neurons in the dorsal horn receive inputs from both skin and internal structures. Although each afferent pathway can be activated independently, they converge on cells in the spinal cord and share a common projection to the brain. Neurons in the brain are however, unable to distinguish between activity arising in one or other of the peripheral pathways, and any activity is assumed to originate in the more commonly stimulated site on the skin.

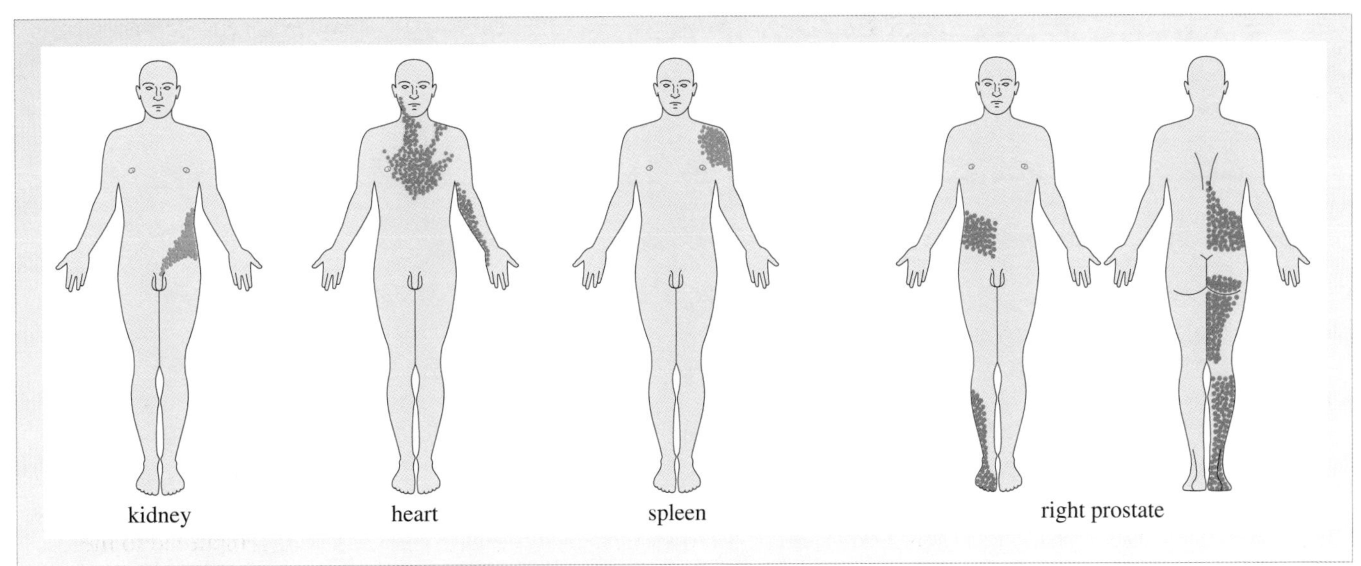

Figure 5.11 Referred pain: examples of pain arising from a visceral disorder referred to a cutaneous region.

silenced using the same analgesics (for example aspirin) as are used to combat pain from peripheral nociceptor activity.

If a wound (e.g. caused by an accident or operation) does not heal quickly and continues to be painful, or if pain continues as a consequence of other conditions (e.g. inflammation or toxins released as a consequence of bacterial or viral attack), then the continual input to the dorsal horn neurons can create a problem known as **wind-up**.

Wind-up is the term used to describe the increasing response given by the dorsal horn neurons to a steady, continuing input from nociceptors. It is as though they have a memory of the nociceptive input and subsequently become sensitized to these inputs and fire more readily. (Sometimes this is called central sensitization to distinguish it clearly from the peripheral sensitization described in the previous section.) The dorsal horn neurons do in fact change in these circumstances. There is an up regulation (cf. down regulation, Block 2. Section 3.2.3) of neurotransmitter receptors and of the production of neurotransmitters. One method of preventing wind-up is to block the NMDA-type glutamate receptors (NMDA stands for N-methyl-D-aspartic acid, see Block 2, Section 3.3.2, Box 3.2) on dorsal horn neurons so they cannot respond to the glutamate released by the nociceptor C fibres. This can be done by procedures such as spinal or epidural injections of an appropriate anaesthetic (Figure 5.12).

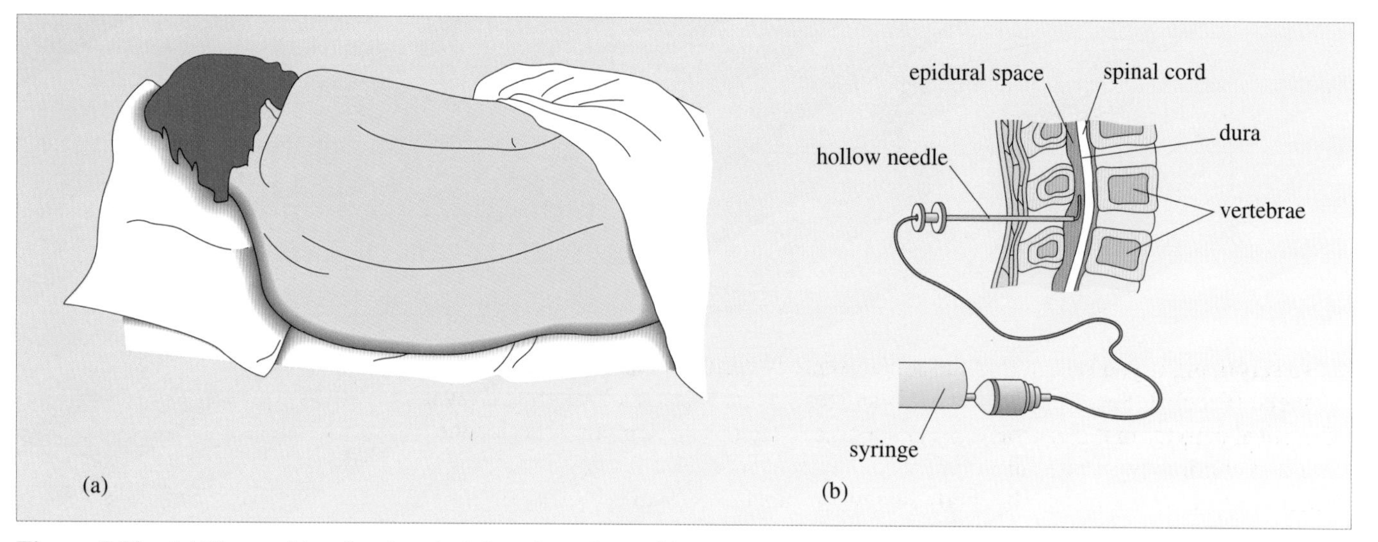

Figure 5.12 (a) The position for the administration of an epidural injection; (b) the position of insertion of epidural anaesthetic.

This procedure is particularly well known for being an effective analgesic during childbirth because it allows the mother to remain conscious when a Caesarean section is necessary (although this is not the only situation when it is used). Its use as an anaesthetic in this circumstance is preferable to a spinal block, where local anaesthetic is injected into the subarachnoid space producing a complete motor block.

❍ In what way is that a disadvantage?

● The recipient of the anaesthetic will not be able to move any part of the body below the spinal block. The anaesthetic is not specific so there will be no messages transmitted in either direction, motor as well as sensory fibres are silenced. An epidural reduces motor function but does not remove it completely.

Spinal or epidural anaesthetic is often given to augment the use of a general anaesthetic when amputation is necessary. Alternatively it may be possible to provide effective local anaesthesia at the site of the amputation itself. Either way the purpose of this procedure is to prevent the dorsal horn neurons from firing during surgery and to reduce the possibility of wind-up. It is hoped this will reduce the incidence of **phantom limb pain** (Figure 5.13).

The majority of amputees (some 60–70%) continue to experience the presence of their missing body part. This is not confined to those who have lost limbs, it includes breasts, genitalia and even internal organs such as the bladder. This may not

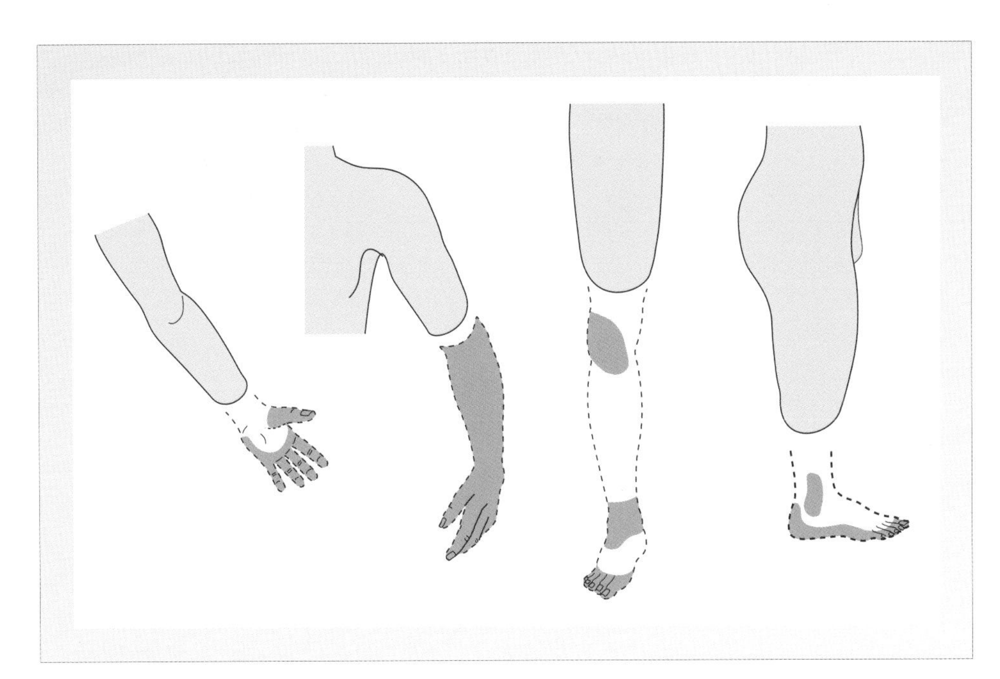

Figure 5.13 Phantom limb experiences of amputees.

be so surprising if you reflect that the somatosensory cortex is topographically arranged (Block 2, Section 2.2.3) and it is suggested that there is a background level of cortical activity in the absence of any specific somatosensory input. It is also the case that children born without limbs frequently possess a clear image of the presence of phantom limbs. This body image appears to reside in the parietal cortex (particularly the right parietal cortex – see Reader Chapter 19). The converse experience is that some people with lesions in that part of the cortex will reject one of their own limbs. They are revulsed by it 'hanging around them' and will try to remove it, pushing it out of bed, even requesting an amputation.

For most amputees, the sensation of the phantom limb lessens as the years pass but initially, there are certain tactile stimuli that are particularly open to misinterpretation. The question below cites one such example.

❍ Using Figure 1 of Chapter 19 in the Reader, explain why a touch to the back of the head might be felt as a touch to a phantom arm.

● Figure 1 shows that the areas of the somatosensory cortex receiving input from the head and from the arm lie adjacent to one another. If the amputee does not see they are being touched it is understandable that they might interpret the signal incorrectly as coming from the much larger area of the cortex that is responsible for receiving sensory input for the missing arm.

It is suggested that the somatosensory cortex gradually reorganizes itself, reducing the area devoted to the missing part, and hence these experiences diminish over time.

Many amputees report pain from the missing area that mirrors the pain experienced prior to the amputation. Originally it was suspected that this was occasioned by sensory input from the stump but it now seems more likely that it should be attributed to wind-up. Hence the current practice of augmenting the general anaesthetic with local analgesia at the site of the amputation so that at least the surgical procedures will not of themselves cause wind-up.

Although phantom limb sensations, which range through the whole gamut of possibilities (cold, hot, wet, sticky, itchy, tingly), diminish over time, pain often does not. This pain can be intractable and is one of the more common causes of chronic pain.

The reasonable hope that chronic pain might be overcome by blocking sensory input has not been borne out. Peripheral analgesics, anaesthetic blocks and even lesioning of sensory and spinal cord pathways have been tried but usually found to give no more than temporary relief. This may be because chronic pain is very often neuropathic in origin, i.e. it is pain experienced as a consequence of malfunction of the nervous system itself. PET scans show that the aversive aspects of pain (the unpleasant emotional experience associated with pain) are generated by activity in the anterior cingulate cortex (Figure 10 in Chapter 21 of the Reader). If activity in this area can be silenced, as has been possible in some experiments using hypnosis (see Chapter 29 of the Reader), then application of a painful stimulus loses its power to cause distress despite being 'felt'. In effect the aim of pain management programmes for those with chronic pain is to tap into the brains ability to police its own activity and to learn to ignore certain messages.

You should now read Chapter 22 of the Reader, *The function and control of pain* by Robin Orchardson.

Activity 5.1 Pain

At this point you may like to study the pain section of the 'Touch and Pain' sequence on *The Senses* CD-ROM. Further instructions are given in the Block 5 *Study File.*

5.6 Summary of Section 5

Pain differs from the other senses in that the perception is unpleasant. It serves as a warning of tissue damage, although some types of damage are difficult to recognize, for example damage to the nervous system that gives rise to neuropathic pain. The double dissociation of pain and tissue damage resulted in beliefs and attitudes to pain and pain control that seem harsh by modern standards. Pain control is of paramount importance nowadays. Since the 1960s when Melzack and Wall developed their influential gate-control theory of pain the quest for effective pain control has focused mainly on neurophysiological studies.

Tissue damage can lead to an inflammatory response and to the release of a cocktail of chemicals that affect nociceptor activity, either directly or (indirectly) by sensitizing the nociceptors. Nociceptors synapse with dorsal horn neurons in the substantia gelatinosa and the afferent pathway ascends in the spinothalamic tract (STT) on the contralateral side of the body. Activity in the STT may be modulated by descending inhibition from the forebrain, and by input to the dorsal horn from the large Aδ peripheral fibres. These ideas form the basis of the gate-control theory of pain relief but the precise details are not yet fully elucidated.

Pain has survival value and is an important diagnostic tool. Knowing about the neurophysiology enables us to make some effective interventions to control pain. However there may be long lasting damage to nociceptors as well as to other neurons in the afferent pathways (e.g. neuropathic damage, central sensitization or wind-up). These may be contributory factors in the experience of chronic pain. Chronic pain, now recognized as a symptom in its own right, can be tackled using a combination of different strategies. The affective component of pain, processed in the limbic system (in the anterior cingulate gyrus) is particularly targeted by many therapies, (i.e. the patient is taught how to ignore the pain).

Question 5.1

Use Figure 5.8 to explain the terms hyperalgesia and allodynia.

Question 5.2

Why do you sometimes feel a sharp, bright first pain but no second pain of the dull, nagging variety?

Question 5.3

Describe the two types of neurons found in the spinothalamic tracts.

Question 5.4

In the light of current knowledge, which descending neurons are more likely to be able to exert a direct analgesic effect in the dorsal horn: serotonergic or noradrenergic? What evidence supports your answer?

Question 5.5

Two people are referred to a pain clinic. One of them, a 32-year-old mother who lost a leg in a road traffic accident, suffers phantom limb pain. The other is a 78-year-old man suffering pain due to cancer which has spread throughout his body and is deemed inoperable. How might the approaches to pain treatment differ in these two cases?

Objectives for Block 5

Now that you have completed this block, you should be able to:

1. Define and use, or recognize definitions and applications of, each of the terms printed in **bold** in the text.
2. Explain the receptor mechanisms involved in the detection and coding of sensory events occurring in the skin. (*Questions 2.2 and 5.2*)
3. Describe the main differences in the organization of neural pathways involved in the transmission to the brain of specific or non-specific sensory information. (*Questions 2.1, 3.2 and 4.1*)
4. Understand the extent to which interaction between senses (and plasticity of the cerebral cortex) can affect our perception. (*Questions 3.1, 3.2 and 4.1*)
5. Distinguish between, and explain the terms body schema and body image. (*Question 3.3*)
6. Explain the suggested basis for endogenous analgesia. (*Question 5.4*)
7. Use knowledge of pathways to comment on drug interventions in pain control. (*Questions 5.3 and 5.4*)
8. Understand and explain some of the problems encountered in the treatment of pain. (*Questions 5.1, 5.4 and 5.5*)

Answers to questions

Question 2.1

Your diagram should resemble Figure 2.10 and show the sensory afferent axon branching in the dorsal horn of the spinal cord. One branch forms part of the dorsal column fibre tract. A second branch synapses once in the dorsal horn and a second order neuron projects via the spinocervical tract to the brain. The third branch gives rise to a local motor reflex loop.

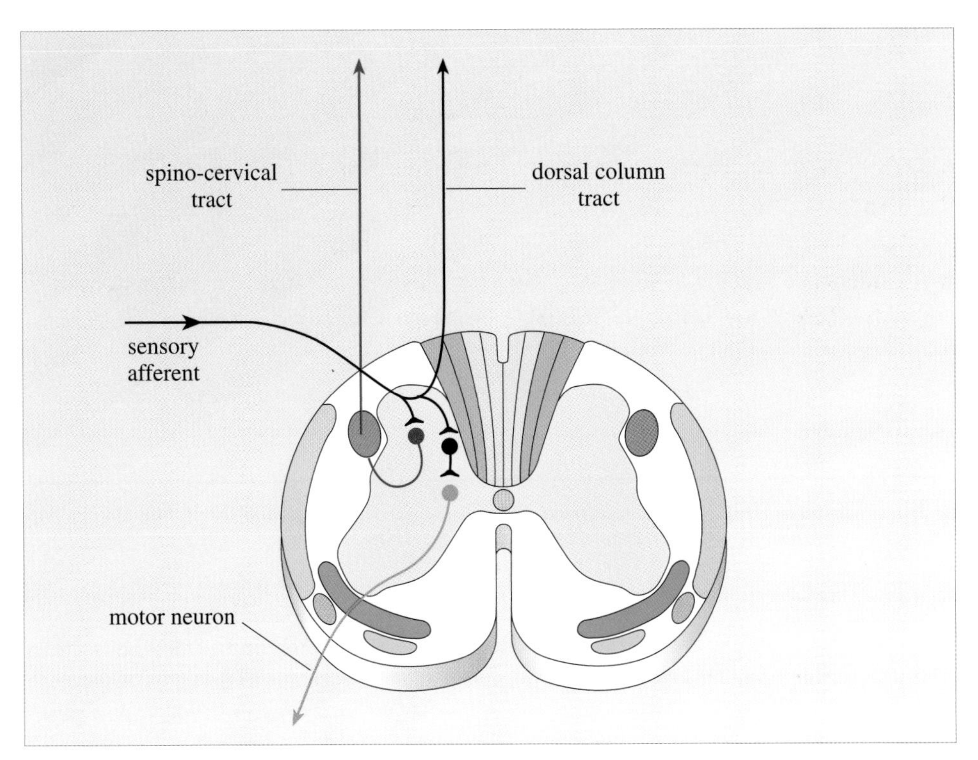

Figure 2.10 Three routes for skin receptor afferents entering the spinal cord.

Question 2.2

Both Meissner's corpuscles and Pacinian corpuscles are rapidly adapting receptors but Meissner's corpuscles have small receptive fields and they are much more numerous (Figure 7, Reader Chapter 18). They, therefore, can function to detect fine detail, as for example when reading Braille, and can also more accurately locate a stimulus on the body surface. The Pacinian corpuscle provides information from a much larger area so does not signal location, although its receptive field has a 'hot spot' just above the Pacinian corpuscle (see Figure 2.11b overleaf). Whilst the Pacinian corpuscle is not sensitive to a single stimulus, it is very responsive to oscillating stimuli (the sensation we describe as vibration) and this is shown in Figure 2.7.

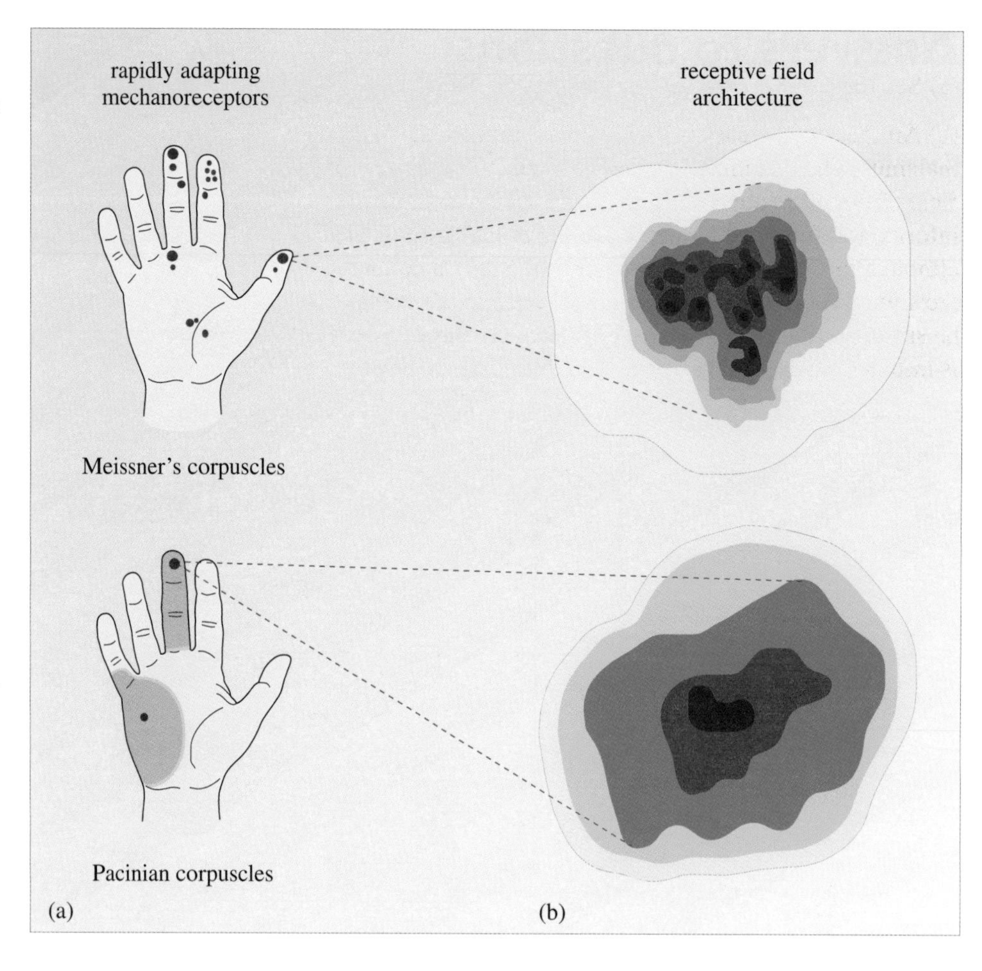

Figure 2.11 (a) Size and structure of receptive fields for Meissner's corpuscles and Pacinian corpuscles. In general Meissner's corpuscles have receptive fields of 2 mm diameter on the fingertips compared to 10 mm diameter on the palm. The Pacinian corpuscle has a much larger receptive field but within it there is an area of maximum sensitivity that is just above the Pacinian corpuscle. (b) Expanded view of the receptive fields. The Meissner's corpuscle has several areas of maximum sensitivity (shown by darker areas) that each overlie the individual corpuscle; one sensory afferent will have 10–25 terminals in the skin, each surrounded by a Meissner's corpuscle. The Pacinian corpuscle has one terminal only but the corpuscle surrounding the terminal is 1–2 mm, which is large compared with other receptor cells.

Question 3.1

Evidence for plasticity of the cerebral cortex is obtained from congenitally blind individuals who are found to use the occipital cortex for successful reading of both Braille and embossed Roman letters (evidence comes from disruption of this process by transcranial magnetic stimulation (TMS)).

Question 3.2

Evidence for crossmodal processing at the cortical level comes from studies of sighted people who find that their tactile discrimination of an object's orientation is disrupted by application of TMS to the occipital cortex. This shows that the visual cortex may process more than just visual inputs. (There is more on crossmodal processing in Block 7.)

Question 3.3

Knowing where all the parts of the car are at any one moment is crucial in avoiding scrapes or worse. So in that sense being able to incorporate the car into a body schema so that one can negotiate small spaces on 'autopilot' is hugely advantageous (and somewhat rebuffs the old adage that if we were meant to drive we would have wheels not legs). Our amazing ability to judge time to contact (TTC), as discussed in Reader Chapter 16, *Motion perception*, supports the idea of it being possible to incorporate a car into our body schema.

Body image is about knowing at a conscious level that we have two arms, even if we cannot feel any sensation from one because we have had a local anaesthetic. As with all perceptions, the body image can be faulty, for example the anorexic individuals who think they are fat. The suggestion in Chapter 19 is that some individuals identify so strongly with their car that it becomes a part of their body image and that they feel themselves to have been physically violated if they return to find that someone has damaged their parked car. (Anecdotally, they are also strongly disinclined to allow anyone else to drive their car.)

Question 4.1

(a) See the diagram below.

(b) All the somatosensory pathways project via the thalamus. The thalamus receives ascending and descending fibres from other areas. The topographically mapped information from the skin is contained in the thalamus and, although the receptive fields of VBN cells in the thalamus are slightly larger than those of the receptor cells, they must be sufficiently informative to guide the motor response that is initiated by the auditory signal.

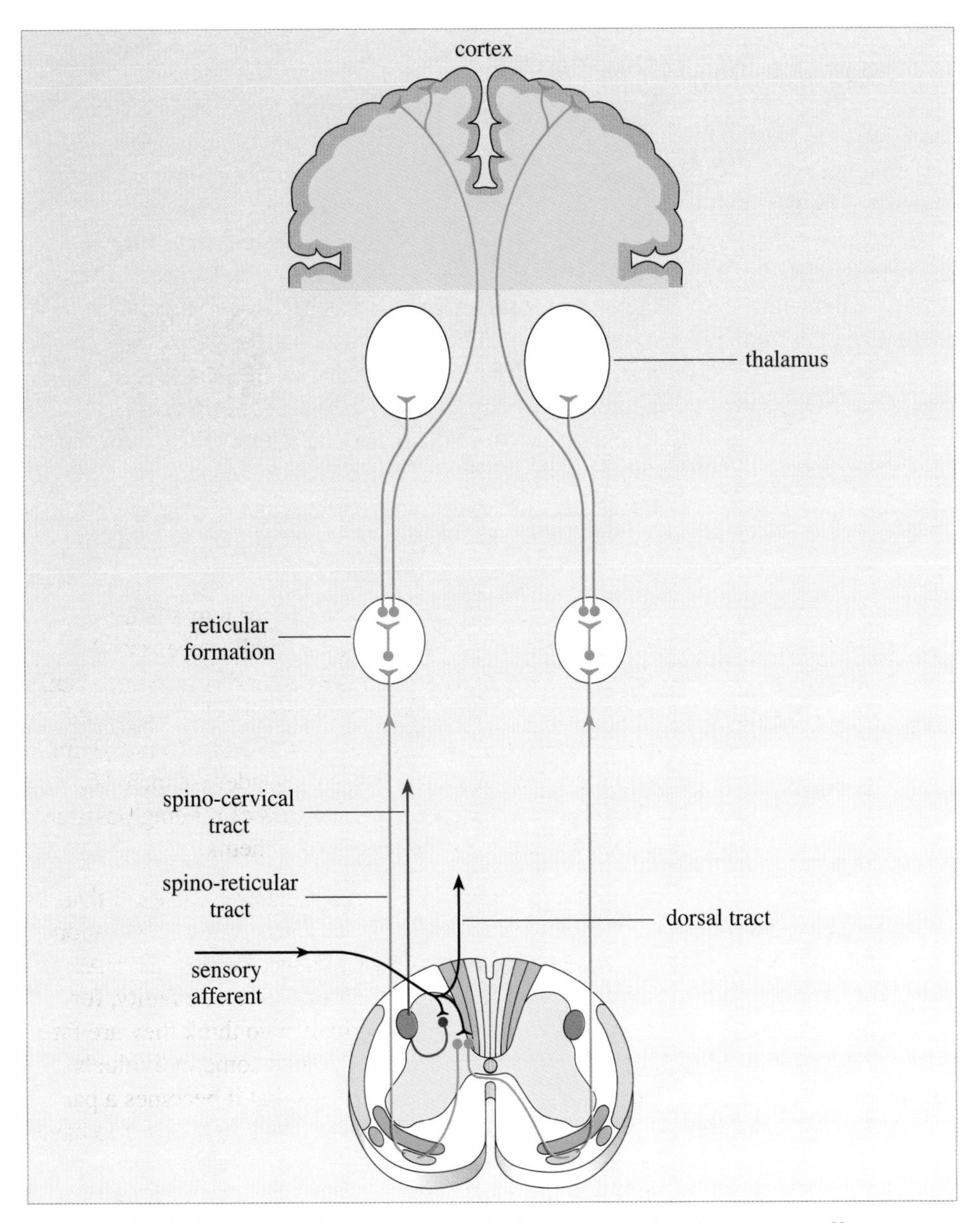

Figure 4.5 A diagram to show the subcortical areas to which the sensory afferents from the right arm project.

Question 5.1

In Figure 5.8 a temperature of 50 °C causes the nociceptor to fire at a rate of about 0.35 s^{-1} initially but if the stimulus of 50 °C is presented again about 4 minutes later the firing rate is about 1.3 s^{-1}. This demonstrates hyperalgesia, which is an increased sensitivity to the same stimulus. The graph also shows that initially a temperature of 40 °C produced no response from the nociceptor but after the skin had been subjected to a temperature stimulus that rose to 60 °C before being removed it will, 4 minutes later, respond to a temperature of 40 °C with a firing rate of about 0.25 s^{-1}. This is allodynia; responding to a previously innocuous stimulus.

Question 5.2

The reason that you sometimes feel first pain but no second pain is that the second pain is mediated by the polymodal nociceptors with C fibres and they are only activated by more intense stimulation. There will be occasions when the stimulus is sufficient to activate a thermo-nociceptor or a mechano-nociceptor but not the polymodal nociceptor.

Question 5.3

There are two types of neuron found in the spinothalamic tracts. The minority are second order neurons that respond only to nociceptive input from Aδ and C fibres terminating in lamina II of the dorsal horn. The majority are wide dynamic range cells with small receptive fields. They receive input from Aδ and C fibre terminations in lamina I but also Aβ fibres in deeper laminae (see Figure 7 in Chapter 20).

Question 5.4

The noradrenergic neurons are more likely to be able to exert a direct analgesic effect in the dorsal horn than the serotonergic neurons. Both terminate in the substantia gelatinosa and when they are stimulated at source (the locus coeruleus and raphe nucleus) exert analgesic properties but noradrenergic agonists are analgesic whereas serotonin agonists are not. So noradrenaline may be directly analgesic but serotonin may be more generally inhibitory and this may be its contribution to the lessening of the perception of pain.

Question 5.5

In each case, the aim is to abolish the pain, or at least reduce it to acceptable levels, while minimizing any side effects. In each case counselling and supportive therapy will be necessary. The mother has a normal life expectancy, and treatment should interfere as little as possible with an undoubtedly busy life. Conservative methods are preferred, such as TENS and local treatment for the stump, aimed at decreasing input to neurons in the dorsal horn. The use of narcotic analgesics or neurosurgery is probably unsuitable for this case. Conservative treatments may be attempted also for the cancer patient, but this is likely to be inadequate. With the shorter life expectancy and greater degree of debility, narcotics may be necessary. Neurosurgery is another possible treatment in this case.

BLOCK SIX

SMELL AND TASTE

Contents

1 Introduction **59**

2 Making sense of smell **61**

2.1 Odorant molecules and their properties 62

2.2 Olfactory receptors 66

2.3 The molecular receptive range of olfactory receptors 69

2.4 Olfactory receptor distribution 71

2.5 Summary of Sections 2.1–2.4 72

2.6 From odorant binding to neural impulse 73

2.7 Combinatorial coding of odours 75

2.8 Odour coding beyond the olfactory neuron 79

2.9 Summary of Sections 2.6–2.8 86

2.10 The effect of concentration on odour perception 87

2.11 Adaptation 89

2.12 Odour discrimination 93

2.13 Summary of Sections 2.10–2.12 106

3 Making sense of taste **109**

3.1 The taste machinery 111

3.2 The molecules of taste 113

3.3 Summary of Sections 3–3.2 116

3.4 Bitter taste 117

3.5 Sweet taste 121

3.6 Umami taste 128

3.7 Summary of Sections 3.4–3.6 132

3.8 Sour taste 133

3.9 Salt taste 139

3.10 Summary of Sections 3.8–3.9 141

3.11 Taste mixtures 141

3.12 The coding of taste 144

3.13 Summary of Sections 3.11–3.12 152

4 **Flavour: a merging of the senses** **155**

4.1 Summary of Section 4 161

Objectives for Block 6 **162**

Answers to questions **163**

Acknowledgements **190**

Glossary for Blocks 5, 6 and 7 **193**

Index for Blocks 5, 6 and 7 **201**

Introduction

1

In *Smelt and tasted*, with all the skill of the poet, W. H. Auden captures the essence of the human capacity for smell and taste:

> The nose and palate never doubt
> Their verdicts on the world without,
> But instantaneously condemn
> Or praise each fact that reaches them:
> Our tastes may change in time, it's true,
> But for the fairer if they do.
>
> Compared with almost any brute,
> Our savouring is less acute,
> But, subtly as they judge, no beast
> Can solve the mystery of a feast,
> Where love is strengthened, hope restored,
> In hearts by chemical accord.

Here we have a succinct summary of some of the main features of these two senses: they are used to extract information about the environment; they are less sensitive than the equivalent senses in other species; they are linked to the emotions; and they respond to chemical stimulation. Auden is correct in associating smell and taste with a chemical origin. Indeed, these two senses are often termed the *chemical senses*. There is something of an irony here. Understanding the world of molecules is commonly held to be difficult, not least because the events that take place at a sub-microscopic level cannot be seen and are explained using abstract concepts. Yet our senses of smell and taste, as you will see, give us a direct interaction with our chemical environment, and our brains respond, apparently instantaneously, to the presence of molecules that can stimulate our smell and taste senses.

Much has been made of the fact that, unlike any of our other senses, those of smell and taste – particularly smell – appear to possess a uniquely evocative quality. Marcel Proust has written of the effect that the smell and taste of a particular type of cake dipped in tea had on his memory of events that had occurred years before. In my own experience, the slightest odour of coal-tar soap transports me back to my childhood and to my grandmother's house. I am sure you have similar experiences.

In evolutionary terms, smell and taste are ancient senses. It seems likely that they developed in primitive organisms as a way of detecting changes to their chemical environment, and in mammals as a means of tracking prey and avoiding predators as well as locating food sources and determining their palatability.

❍ If that is so, take a minute to jot down why we have developed two complementary systems for sampling our chemical environment; that is, from what you know about smell and taste try to identify any differences between how the two senses operate.

● Our ability to smell allows us to detect materials in the air. These materials tend to be volatile – for example perfume, or the fragrance of a flower – and because our noses do not have to be in contact with the source of the odour, they can be detected at a distance (in some instances hundreds of metres away). Our ability to smell seems to help us identify whether we should be attracted to or repelled by whatever is the source of the **odour**. Our ability to taste, however, requires us to put the materials into our mouths, so we need to be in direct contact with them. Taste is not something that can be experienced at a distance. In most cases, taste detects substances present in the solids and liquids we require as foods and informs us as to whether or not these are acceptable to eat. And rather than being volatile, these substances often have to be soluble in water to some extent.

How is it, then, that molecules are capable of producing the sensory triggers that are necessary for us to smell and taste our world? To understand how answers to this question are formulated you need to have an understanding of how molecules are constructed and how they interact. That is precisely where we start in our description of the sense of smell.

Making sense of smell

2

We are well aware that our sense of smell, or **olfaction** (from the Latin *olfacere*, to smell), resides in our nose. While the anatomy of smell will be dealt with in somewhat more detail in the Reader chapters, as well as on *The Senses* CD-ROM, it is sufficient for now to recognize that, at the top of the nasal passage, there is a $5\ \text{cm}^2$ patch of epithelial cells, called the **olfactory epithelium** (Figure 2.1).

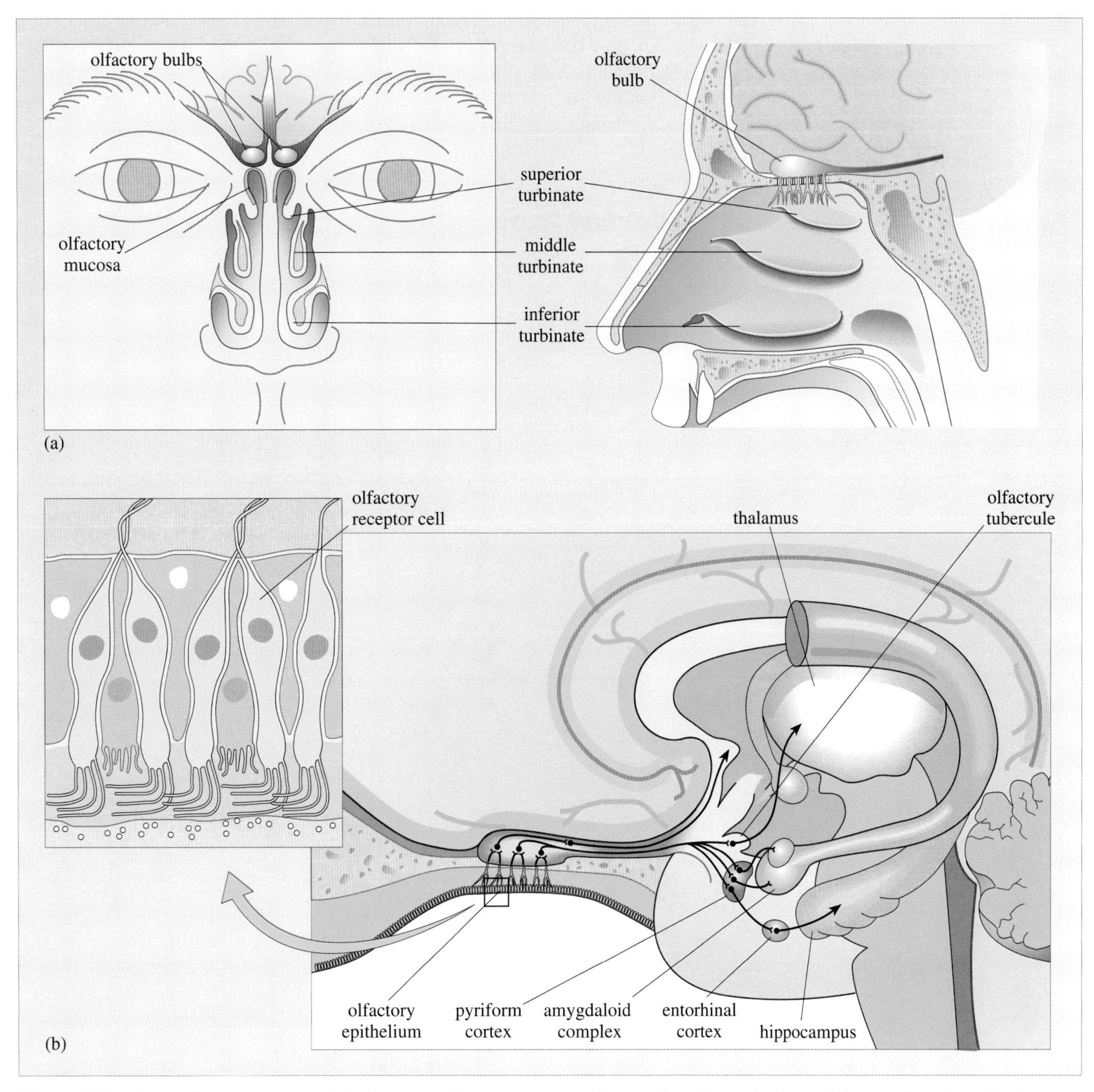

Figure 2.1 The main components of the human olfactory system: (a) rostral and lateral views of the nasal cavity; (b) the olfactory apparatus and connections to the brain, with olfactory receptor cells shown in the enlarged detail.

This contains the nerve cells, the **olfactory neurons**, which respond to the odours that waft up into the nasal passage. These olfactory neurons initiate and conduct electrical impulses to the **olfactory bulb** that is a specialized part of the brain. The bulb then processes these input signals for onward transmission to the olfactory cortex and from here to other parts of the brain. Using these pathways, odour signals reach the regions involved in the conscious perception of odours as well as the **amygdala** and hypothalamus, areas of the limbic system involved in emotion and motivation.

Consequently, our study of the sense of smell begins with the properties of the molecules that give rise to an odour, progresses to an understanding of how these molecules interact with the olfactory neurons, subsequently develops the current thinking about how the neural impulses carry odour information to the brain, and finally discusses the ability of the system to recognize and discriminate between smells.

2.1 Odorant molecules and their properties

The odorant molecules, or, more simply, **odorants**, responsible for the smells we sense are small organic molecules of relative molecular mass no larger than about 300. This immediately raises the question, 'How many atoms are there in a molecule of mass 300?' The following calculation attempts to address this issue.

❍ The general molecular formula of one of the very simplest classes of organic compounds, the hydrocarbons (of which the fuels methane and butane are members), is C_nH_{2n+2}. Here n can be any whole number > 0, so when $n = 1$ the formula of the molecule is CH_4 (methane) and when $n = 4$ it is C_4H_{10} (butane). Given that the relative masses of carbon and hydrogen atoms are 12 and 1, respectively, use the above general formula to calculate the approximate number of carbon atoms, n, in a molecule of mass 300.

● If the molecule contains n carbon atoms, they must contribute $12n$ mass units. Similarly, the $2n + 2$ hydrogen atoms must contribute $2n + 2$ mass units. So, we can write $12n + 2n + 2 = 300$,

or $14n + 2 = 300$,

or $14n = 298$.

This means that $n = 298/14 \approx 21$.

Consequently, odorants are small organic molecules containing no more than about 20 carbon atoms. If they also contain nitrogen (relative mass = 14), oxygen (relative mass = 16) or sulfur (relative mass = 32) atoms, then, of necessity, they will have to contain fewer carbon atoms than this.

In general, molecules of larger mass are ineffective odorants because they have insufficient volatility. It is possible to appreciate this from the boiling temperatures at one atmosphere pressure of three simple ester odorants: ethyl formate, which contains five 'heavy' atoms (three carbon, two oxygen); ethyl nonanoate, which contains thirteen such atoms; and ethyl palmitate which contains twenty (Table 2.1). The boiling temperature of any substance is the temperature at which the pressure exerted by the gaseous molecules of that substance above its liquid phase is equal to the pressure of the surrounding atmosphere. At any given pressure, the more **volatile** substances are those with the lower boiling temperatures. This is because, for these substances, there are more molecules in the gaseous phase at lower temperatures – that is, they are easily vaporized – and the more gaseous molecules there are, the higher the vapour pressure.

Table 2.1 The boiling temperatures (b.t.) at one atmosphere pressure of three ester odorants.

Odorant	Chemical formula	Mass	Odour	b.t./°C
ethyl formate	$C_3H_6O_2$	74	strong, fruity, rum-like	54
ethyl nonanoate	$C_{11}H_{22}O_2$	186	cognac, nut-like	227
ethyl palmitate	$C_{18}H_{36}O_2$	284	faint, waxy-like	340

❍ From the boiling temperature data at one atmosphere pressure what can you say about the relative volatility of the three ester odorants?

● Ethyl formate is much more volatile than the other two esters; it boils at a temperature well below the boiling temperature of water (100 °C). Ethyl nonanoate boils at a temperature well above that of water and so is less volatile, and ethyl palmitate boils at a temperature that is even higher so it is not very volatile at all.

As the volatility decreases, the odour changes in two ways. First, the quality of the odour varies from fruity, through nut-like to waxy. We shall discover later why such differences occur. Second, the intensity varies with molecular size. The smaller, more volatile, molecules give rise to stronger, more intense odours, whereas the larger, involatile molecules produce only faint odour perceptions.

Volatility, then, is a requisite property of an odorant. But how is it that volatile molecules are capable of producing the sensory triggers that are necessary for us to smell our world? In vision and hearing, energy in the form of light or pressure waves is transduced into an electrical signal that is communicated to the brain. How is a molecule capable of interacting with the smell detection system housed in our nose in such a way that it produces an electrical signal that can be interpreted by our brain? To understand the answers to this question you first need to have an understanding of how molecules are constructed and how they interact. Box 2.1 (overleaf) contains a brief outline of the variety observed in chemical bonding. It is not something you will be required to remember; rather, it is presented here as an aid to help your understanding of later material.

Take a look at Figure 2.2 (overleaf), which contains the structures (here shown as skeletal diagrams) of some representative molecules that have well-known smells.

❍ Compare the masses and the structures (that is, the patterns of bonding and molecular shapes) of benzaldehyde (C_7H_6O) and hydrogen cyanide (HCN), both of which have an almond smell. What molecular features do they have in common?

● Benzaldehyde has a mass of 106 whereas hydrogen cyanide has a mass of 27. Not only are the masses very different, so are the bonding arrangements and molecular shapes. Benzaldehyde contains a ring of carbon atoms and a C=O

Box 2.1 Variety in covalent chemical bonding

All molecules are built up from atoms, the most common in the organic molecules involved in smell being those of carbon, hydrogen, oxygen, nitrogen and, to a lesser extent, sulfur. For each chemical compound the precise arrangement of atoms is unique; all molecules of the same chemical are identical, but molecules of different substances differ in the ways in which their atoms are arranged. These different structures are what give the molecules their differing physical and chemical properties. The structure of a molecule comes about because the atoms in it are held together by **covalent bonds**. These are bonds that are formed by two atoms sharing at least two (sometimes four or six depending on the atoms involved) electrons. When written on the page we represent a two-electron covalent bond by a '—' between the atoms that are bonded together, for example, C—H for a bond between carbon and hydrogen. It turns out (and we haven't space here to go into the reasons why) that hydrogen can only form one such covalent bond, oxygen can form two, nitrogen three and carbon four; sulfur can form two, four or six. This means that, even with this simple array of atoms and bonding possibilities, for the odorant molecules that contain up to twenty carbon atoms there are essentially millions of different arrangements for the atoms to bond and therefore millions of possible compounds that can exist.

However, there is even greater variety than this suggests. Consider an oxygen atom: this atom could form two covalent bonds to two different carbon atoms, C—O—C, or it could form two covalent bonds to the same carbon atom, C=O. The latter is called a double bond and the carbon and oxygen atoms are sharing four electrons (two in each bond). These two bonding arrangements are different so the compounds that contain them will have different properties.

It is also possible for nitrogen atoms to form double bonds, for example, C=N, as well as triple bonds, C≡N. So, despite odorant molecules being limited to molecular masses less than 300, you can begin to appreciate that there is enormous potential for molecular diversity.

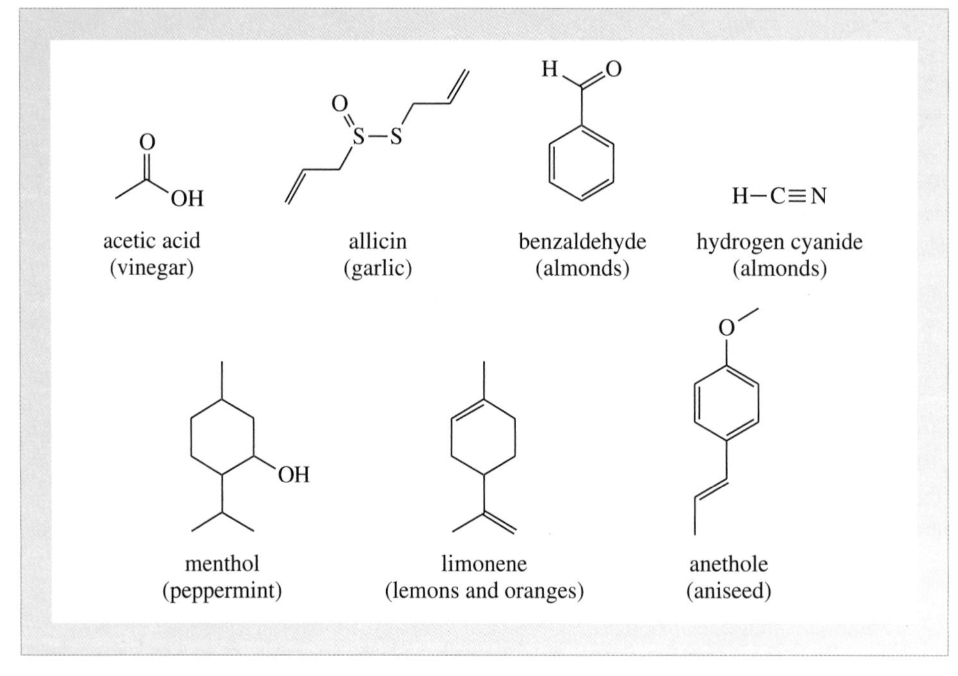

Figure 2.2 The molecular structures of some common odorant molecules.

group whereas hydrogen cyanide is linear and contains a C≡N group. There appears to be no common feature, other than the C—H bond, that could be considered to give rise to an almond smell.

In fact, all the other molecules in Figure 2.2 contain C—H bonds, and they don't elicit an almond smell so it is unlikely that this feature is responsible for an almond odour.

❍ Now compare menthol and limonene. What can be said about their structures and the odours they elicit?

● Menthol and limonene have quite similar structures. They both contain the same carbon skeleton, including a six-carbon ring system, and two of the groups attached to the ring are very similar in size and bonding arrangement. The only major differences are the presence of C=C double bonds in limonene and the C—O—H in menthol. So small changes in chemical structure can give rise to significant changes in the smell that is perceived.

Indeed, any theory of how the brain interprets smells must account for these two observations: *chemically dissimilar molecules can give rise to similar odours, and chemically similar molecules can give rise to dissimilar odours*.

The covalent bonds that hold the atoms in a molecule together are very strong and require a significant input of energy to break them. These bonds only break during chemical reactions, when a molecule is transformed permanently into a separate and completely different molecule. These are not the kinds of changes that are associated with a physical property of a substance, such as its smell. However, when two *different* types of atom are bonded together, such as carbon and oxygen, the electrons they share are not shared equally; the electrons spend more of their time nearer to the oxygen atom (this is not the place to explain why this is so). As electrons are negatively charged, this means that, in relative terms, the oxygen atom has a slight excess of negative charge associated with it.

❍ What effect do you think this will have on the carbon atom bonded to the oxygen atom?

● The carbon atom will have a slight deficiency of negative charge so, in relative terms, it will be slightly positively charged.

It turns out that carbon and hydrogen atoms have roughly similar propensities to attract electrons, nitrogen has a stronger propensity than either, and oxygen stronger still. So we can think of these bonds as shown in Figure 2.3, where the symbols δ+ and δ–, pronounced delta plus and delta minus, respectively, mean slight positive and slight negative charges. The bonds are said to be polarized. Note that C—H and C—C bonds are essentially not polarized.

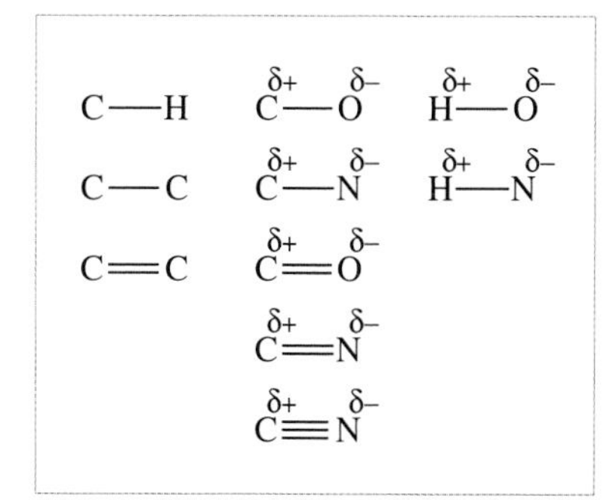

Figure 2.3 The polarization of covalent bonds.

These polarized bonds have a profound effect on the ways in which molecules can interact. For example, the slightly negatively charged oxygen and nitrogen atoms of one part of the molecule can be attracted to a slightly positively charged centre elsewhere in the same molecule or even in a different molecule. These are called polar attractions. The most common type of polar interaction is with a hydrogen atom that is itself bonded to oxygen or nitrogen. This is the kind of interaction that is present between two water molecules (Figure 2.4), for example, and is called **hydrogen bonding**. We depict the hydrogen bond interaction by a dashed line.

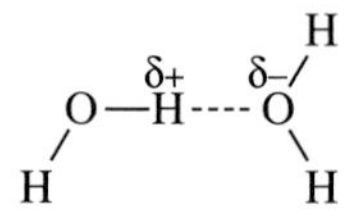

Figure 2.4 Hydrogen bonding between two water molecules.

In terms of energy, hydrogen bonding is much weaker than covalent bonding; in fact, a hydrogen bond is about one-tenth as strong. This makes hydrogen bonding useful for temporary interactions between molecules because they can easily be made and can easily be broken. Of course, the more hydrogen bond interactions that there are between two molecules the stronger will be the overall interaction.

There is one other type of weak interaction between molecules that contributes to how molecules give rise to odour. This interaction is caused by **London forces**. If you take a look at the structure of limonene, which is one of the molecules that give rise to the smell of lemons, you should notice that it only contains carbon and hydrogen atoms. Now, we've already noted how C—C and C—H bonds are not polarized to any great extent. We should qualify this statement: *on average* these bonds are not polarized. However, the electrons in these bonds are in a constant state of motion and at any one instant in time may be nearer to the carbon atom than the hydrogen atom (or vice versa). This results in a small *transient* polarization of charge. Any transient regions of negative charge in one molecule will create transient areas of positive charge in a neighbouring molecule These areas of transient opposite charges attract one another. Averaged out over a period of time, this gives rise to the net attractive forces that we call London forces. Not surprisingly, the strength of London forces is very weak, about one-tenth the strength of a hydrogen bond. However, when there are many C—C and C—H bonds in a molecule, the *overall* contribution of London forces to the total interaction between two molecules can be considerable.

One important outcome of the above discussion about hydrogen bonding and London forces is that it is possible to predict which parts of one molecule will interact with which parts of a second molecule, namely:

- non-polar regions, i.e. regions rich in C—C and C—H bonds, of one molecule will interact with a similar region in another molecule;
- regions that contain C—O, C—N, O—H and N—H bonds, will interact with similar regions in another molecule.

So, now we have an appreciation of how molecules are able to interact, but how does the nose use this molecular interaction to smell an odour? To make sense of that, we need to develop an understanding of the molecules that the odorants interact with, the olfactory receptors.

2.2 Olfactory receptors

Through evolutionary development, all living creatures have the means by which their cells can respond to the presence of chemicals in the external environment (external here means outside the cell, rather than outside the organism). You may remember from Block 2 that, to effect this, each cell has embedded in its membrane specialized molecules, called receptor proteins, which respond in very specific ways when molecules interact with them. Such specialized receptor proteins are distributed throughout the body and they are responsible, amongst other things, for how we respond to hormones and for chemical and electrical communication in the nervous system. Indeed many drugs exert their pharmaceutical effect by binding to particular receptor proteins.

The smell receptor proteins, called **olfactory receptors**, are, by comparison to the odorant molecules they interact with, very large. The olfactory receptors are made up of some 320 or so amino acid units linked together and therefore have relative molecular masses in the region of 40 000–50 000 – about one hundred times the size of an odorant molecule. These receptors are embedded in the cell membrane in the cilia, finger-like projections that are found on the surface of the olfactory neuron. The cilia from all the neurons form a mat on the surface of the olfactory epithelium and are coated with a layer of mucus. It is in this matrix that the interaction between the odorant molecules and the olfactory receptors takes place.

Like all proteins, olfactory receptors have a carbon (C) and a nitrogen (N) terminus. The C-terminus of the olfactory receptors is located on the inside of the cell; the protein then traverses the membrane seven times (just like the protein rhodopsin involved in vision) to leave the N-terminus in the extracellular matrix on the outside of the cell (Figure 2.5).

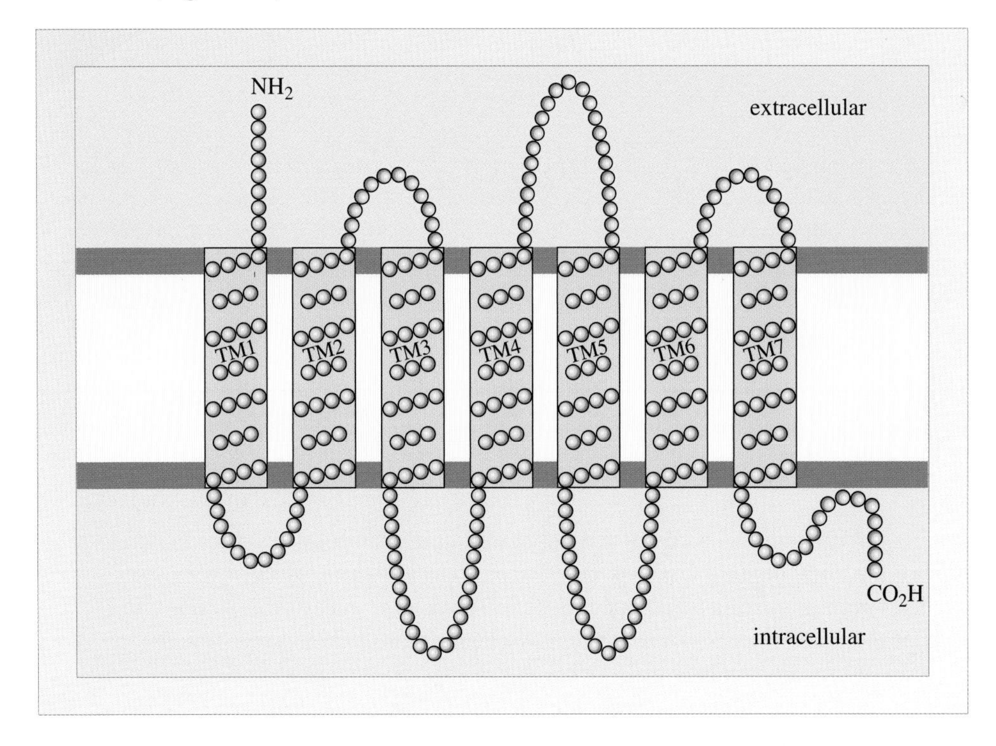

Figure 2.5 The arrangement of the olfactory receptor in the cell membrane. The seven membrane-spanning segments are labelled TM1–TM7 (see later).

In general, because they are embedded in cell membranes, the structures of proteins like the olfactory receptors are much more difficult to study than proteins present in the cytoplasm. One crucial reason for this is that they can't be crystallized, so the precise position of the atoms cannot be studied by X-ray crystallography. Despite this drawback, by comparing their amino acid sequences with those of other proteins whose structures are known, it is possible to predict, for example, which segments will adopt helical arrangements and which will be the amino acids that make up the membrane-spanning segments.

❍ Given that the membrane environment is a very non-polar one, what kinds of amino acids do you think will be the ones involved in forming the membrane-spanning segments?

● Since a non-polar region of one molecule will interact preferentially with the non-polar region of another, the amino acid residues that span the membrane must be non-polar.

With the advent of modern computers, provided the amino acid sequence of a protein is known, it is now possible to use the above approach to predict the probable organization of the protein when it is embedded in a membrane. This has been done for a rat olfactory receptor, called OR-I7. The amino acid residues that make up the membrane-spanning segments of the protein, called transmembrane (TM) domains, are organized into helical structures. The relative orientation of these domains is shown in Figure 2.6.

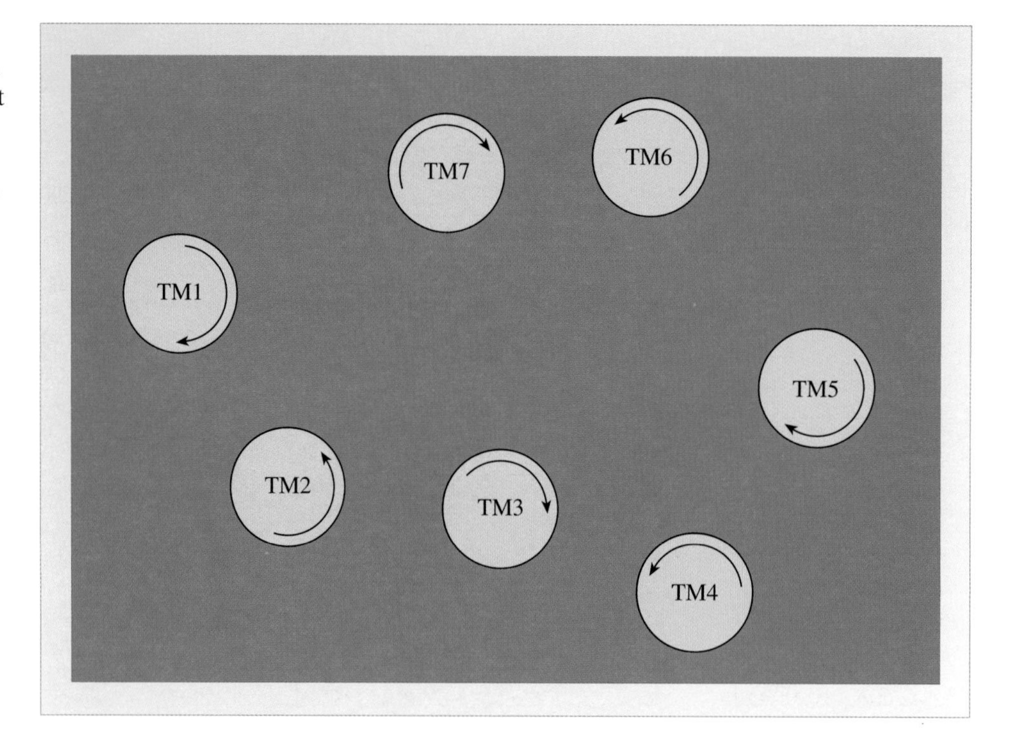

Figure 2.6 A cross-section, viewed from the extracellular side, of the seven TM domains of the rat olfactory receptor OR-I7. (Arrows indicate direction of helices from the N-terminus to the C-terminus.)

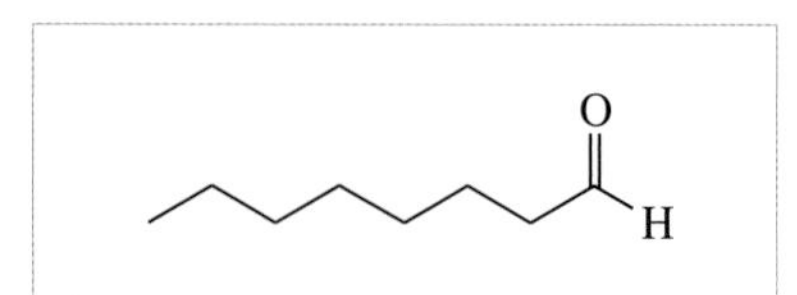

Figure 2.7 The structure of octanal, the odorant that binds to olfactory receptor I7.

In general, the outer surfaces of the helices contain non-polar residues that help to anchor the domains in the non-polar environment of the membrane; the inner surfaces contain residues that are capable of binding to polar groups. For OR-I7, four of the TM domains, TM4–TM7, have been found to be responsible for the binding of the odorant molecule, octanal (Figure 2.7), one of the odorants recognized by this olfactory receptor.

❍ Examine the structure of octanal. What kinds of molecular interaction do you think will occur between it and the olfactory receptor?

● Octanal contains a long, non-polar, hydrocarbon region that is likely to interact with a similar region in the receptor protein via London forces. Octanal also contains a polarized C=O group, which could form a hydrogen bond to an N—H or O—H group on the protein.

This is precisely what is found. Within the olfactory receptor, about 1 nm below the extracellular surface of the membrane (about one-eighth of the way in) there is an octanal-binding pocket formed by the TM domains 4–7. Seven amino acid residues in this pocket contribute to the binding of octanal via non-covalent interactions. The strongest binding interaction is, not surprisingly, a hydrogen bond between the oxygen atom of octanal and a hydrogen atom of a positively charged ammonium group of a lysine residue, positioned at amino acid residue 164 (Lys164) in the

TM4 section of the olfactory receptor. This is stabilized by a polar attraction of the carboxylate group of Asp204 on TM5. The remaining binding interactions involve London forces between the octanal hydrocarbon backbone and amino acid hydrocarbon side chains. The most important of these are those involving phenylalanine (Phe205) and alanine (Ala208) on TM5, alanine (Ala258) and phenylalanine (Phe262) on TM6, and valine (Val281) on TM7. These interactions are shown schematically in Figure 2.8.

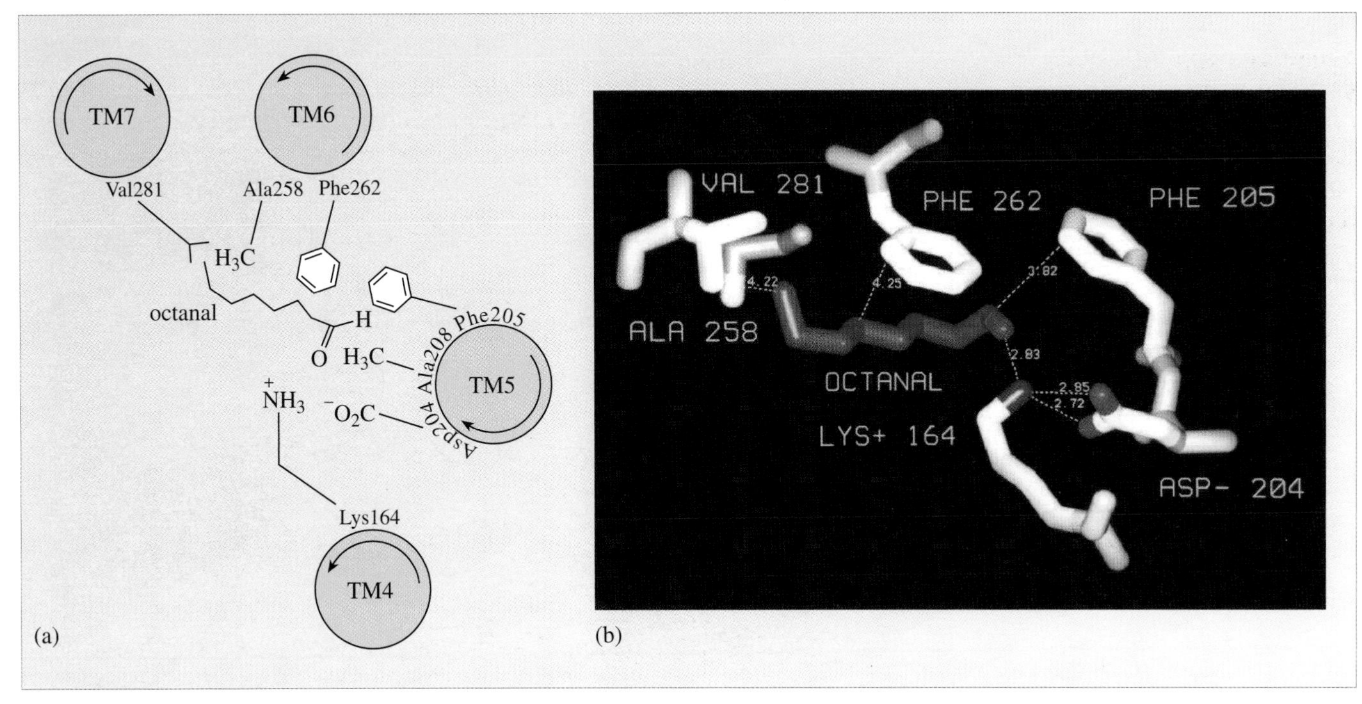

Figure 2.8 Binding interactions between octanal and the rat olfactory receptor OR-I7: (a) cross-sectional view; (b) a three-dimensional view.

2.3 The molecular receptive range of olfactory receptors

What makes this model of how the odorant molecule binds to the olfactory receptor so compelling is that the computer-calculated binding affinities for a series of related molecules correlate well with the electrical activity in the olfactory epithelium as measured experimentally (Table 2.2 overleaf). For example, for the series of aldehydes hexanal to undecanal, each of which differs from the previous member of the series by only one CH_2 group, the computer model predicts that optimal binding is found with octanal, which is the compound that exhibits maximal olfactory activity. Moreover, binding affinity and olfactory activity are seen only for those aldehydes that have carbon chain lengths similar to that of octanal. The receptor therefore can be seen to be selective for a particular range of carbon chain lengths.

What is also very interesting is the selectivity, both in olfactory receptor binding affinity and in olfactory activity, for molecules containing the aldehyde functional group, H—C=O. If you compare the data for octanal with those for octanol, octanoic acid and octane, all of which contain eight carbon atoms, you can see that only octanal has any biological activity and it is also the molecule with highest affinity for the olfactory receptor.

Table 2.2 Comparison of OR-I7 predicted binding affinity (relative to octanal = 1) with electrical activity in the rat olfactory epithelium of a series of compounds related to octanal.

Odorant molecule	Predicted affinity	Increase in electrical activity (%)
hexanal	0.004	0
heptanal	0.04	35
octanal	1	72
nonanal	0.3	51
decanal	0.03	43
undecanal	<0.0001	0
octanol	0.006	0
octanoic acid	0.07	0
octane	<0.0001	0

Thus, it would appear that the OR-I7 olfactory receptor is narrowly tuned to its ligand, octanal, and closely related molecules. The range of structures that the olfactory receptor is responsive to is called its **molecular receptive range**.

Similar observations have been made for a mouse olfactory receptor, OR-S25, which is finely tuned to respond to the alcohols hexanol and heptanol. In contrast to the rat OR-I7 receptor, this receptor employs fifteen amino acid residues on domains TM3–TM7 to bind these molecules. Once again, a lysine residue, Lys302 on TM7, is involved in a crucial hydrogen bond to the alcohol oxygen atom. It seems most likely that, despite the fine differences between the two receptors, this type of binding of an odorant to an olfactory receptor is a universal mechanism in the way all mammals, including humans, smell their environment.

Activity 2.1 Olfactory molecule binding

Now would be a good time to work through this activity on the CD-ROM, in which you can examine further the interaction between the odorant hexanol and the mouse olfactory receptor, S25. Further details are given in the Block 6 *Study File*.

2.4 Olfactory receptor distribution

So far, we have discussed the binding of an odorant molecule to an olfactory receptor as if the nose contained only one such receptor. Of course, this cannot be so, otherwise all molecules that bound to it would smell the same. What is more, it is clear that both the rat, I7, and mouse, S25, olfactory receptors, bind only a narrow range of molecules. If the same were true for humans then, if we had only one type of olfactory receptor protein, there would be a very limited range of odorants that we could detect. Clearly, there must be more than one receptor, but how many more?

An early theory, proposed by the American chemist John Amoore in the 1960s, suggested there was a small group of receptors that could be assigned to what were considered primary odours, namely camphoraceous, musky, floral, pepperminty, ethereal, pungent, and putrid. By considering the three-dimensional molecular structures of hundreds of odorants, Amoore concluded that each of these receptors had a different, and specific, shape (Figure 2.9). The theory proposed that if an odorant molecule had a particular size and shape that enabled it to fit into one of these receptor sites it would elicit that particular odour. If its molecular dimensions allowed it to fit into more than one site, then the odour would be a composite of the two primaries. This theory proved attractive because, first, it is simple, and second, it is similar to the concept of primary colours. However, we already know that a very small molecule like HCN elicits a smell that is the same as that from the larger benzaldehyde. Moreover, simplicity is not a prerequisite of an idea being correct!

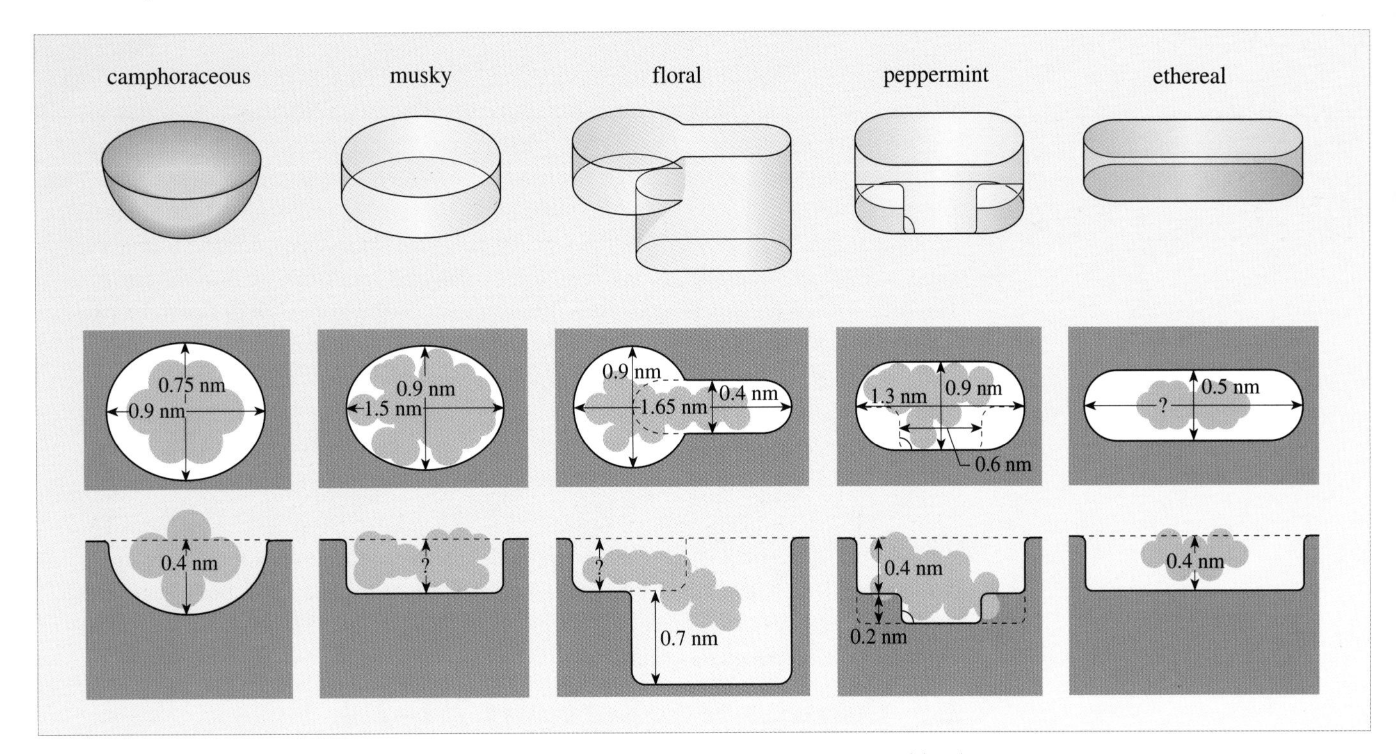

Figure 2.9 The shapes and dimensions of some of the primary receptors proposed by Amoore.

In fact, there is a far greater number of different receptors, about 350 in humans, and how we use these to interpret a smell is far more complex than initially thought. Even so, a refined version of Amoore's theory is still used by the perfumery industry, where molecules are synthesised with particular shapes and with particular groups of bonding patterns, called **olfactophores**, with the aim of generating new molecules that will contribute a particular odour to a fragrance. For example, you might consider the aldehyde group and the hydrocarbon chain as the two olfactophores that contribute to the odour of octanal. Of course, these groups do not have any odour in their own right; rather, they are the groups that enable octanal to bind to one or more olfactory receptors.

Using the techniques of molecular biology, together with some inspired guesswork, the American scientists Linda Buck and Richard Axel discovered in the early 1990s that mammals possessed genes for approximately 1000 different olfactory receptors. In humans, though, it was later found that only about a third of them are expressed (that is, the olfactory receptor protein for which the gene codes is synthesised). Amoore's earlier model was clearly a gross simplification of reality! Moreover, although there are millions of olfactory neurons in the olfactory epithelium, it turns out that each gene is expressed in 1 in every 1000 neurons.

❍ If there are about 1000 different olfactory receptors, what does this pattern of gene expression tell you about the number of receptors in each olfactory neuron?

● If there are 1000 genes and these are expressed in 1 in every 1000 neurons there must be one olfactory receptor in each olfactory neuron.

This means that each olfactory neuron only responds to the odorant molecular receptive range of the olfactory receptor it contains.

Before we continue, it is worth considering how the neurons that contain the same type of olfactory receptor are distributed in the olfactory epithelium. Are they all housed in the same region, or are they distributed randomly across the epithelium? In humans we still don't know the answer to this question, but in rodents it has been found that the olfactory epithelium appears to be divided into four zones of similar area. Most neurons that contain the same type of olfactory receptor are found in the same zone of the epithelium, but they are distributed randomly within the zone (Figure 2.10). Why this is so is currently unclear, as the zones do not correlate with any regional preferences for odorant sensitivity, with development or with the distribution of the genes in the **genome**. One possible explanation is that it has the effect of simplifying the 'neuronal wiring'. The axons from the olfactory neurons project through to features in the olfactory bulb called glomeruli. Since any one glomerulus only receives axons from neurons that contain the same type of olfactory receptor, housing these neurons in the same zone means the pathways the axons must take are simplified.

You should now read Chapter 23 of the Reader, *The olfactory sensory system* by Tim Jacob.

2.5 Summary of Sections 2.1–2.4

The sense of smell detects small, volatile odorant molecules. These interact with olfactory receptor proteins that are embedded in the membrane of the cilia of the olfactory neurons which are housed in the olfactory epithelium. The interactions between odorant and olfactory receptor are chemical in nature and involve hydrogen bonding and polar attractions, via the polar functional groups of the odorant and the receptor, and London forces between their non-polar hydrocarbon regions.

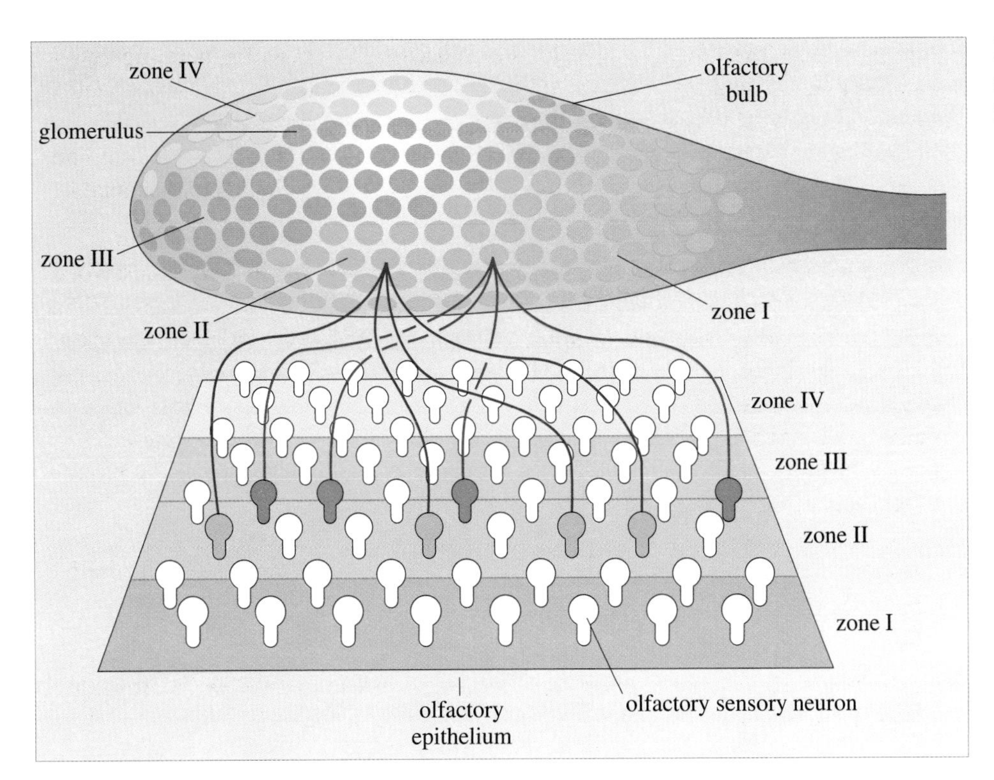

Figure 2.10 The zonal distribution of olfactory neurons in the rodent olfactory bulb.

Humans possess genes for about one thousand different types of olfactory receptor, but only about 350 are expressed. The olfactory neurons of rodents are organized into four zones, each type being found in one zone. Each olfactory neuron appears to express only one olfactory receptor and each type of receptor possesses a molecular receptive range, that is, a range of odorant molecular structures that can bind to the receptor. These seem to be narrowly tuned to consecutive members of molecular series.

2.6 From odorant binding to neural impulse

While the binding of odorant molecules to olfactory receptor proteins allows the olfactory system to *detect* odours, there must be a means by which this chemical information is *transduced* into the electrical signals that are transmitted to the brain. In brief, this is achieved by coupling the odorant binding process in the cell membrane to a series of chemical reactions inside the cell that ultimately bring about a depolarization of the cell membrane potential from –65 mV to about –45 mV. This depolarization gives rise to action potentials, which constitute the electrical signal that is transmitted along the axon to the glomeruli in the olfactory bulb. Here the bulb processes the electrical signals from all the activated neurons for onward transmission to the olfactory centres in the brain.

How is cell depolarization achieved? Well, on the interior surface of the cell membrane the olfactory receptors are intimately connected to a protein called a G-protein. When the odorant binds to the olfactory receptor, the G-protein is released and this migrates within the cell to activate an enzyme, adenylyl cyclase, that synthesises a molecule called cyclic AMP (cAMP) from ATP. The cAMP binds to and opens an ion channel that allows the positively charged ions Na^+ and Ca^{2+} to enter the cell. It is this influx of positive charge that depolarizes the cell. In fact, the increase in concentration of Ca^{2+} within the cell results in the opening of a second ion channel that allows the negatively charged ion Cl^- to exit the cell. The loss of the negatively-charged ions from within the cell also contributes to membrane depolarization. These processes are summarized in Figure 2.11.

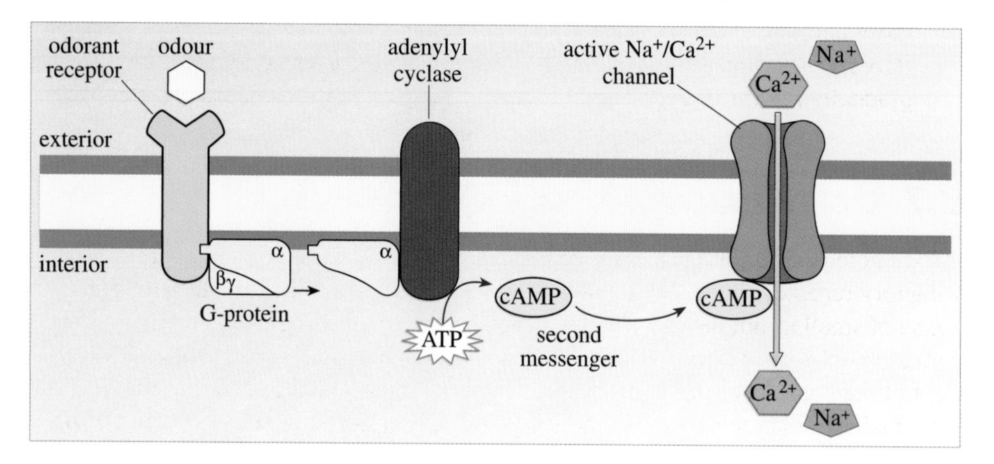

Figure 2.11 The steps in the production of olfactory neuron depolarization.

This change in the Ca^{2+} concentration within the olfactory neuron can be monitored by a fluorescence technique. To achieve this, olfactory neurons are exposed to a substance, called fura-2, that can bind specifically to Ca^{2+}. This compound is fluorescent; that is, it emits light at a wavelength of 510 nm when it is exposed to light of wavelength 380 nm. However, when fura-2 is bound to Ca^{2+} the fluorescence only occurs when exposed to light of wavelength 340 nm. So, when olfactory neurons that are loaded with fura-2 are exposed to odorants, the Ca^{2+} flux into the cell results in a reduction in the concentration of 'free' fura-2. The outcome is a reduction in the fluorescence from exposure to 380 nm light, as you can see from Figure 2.12.

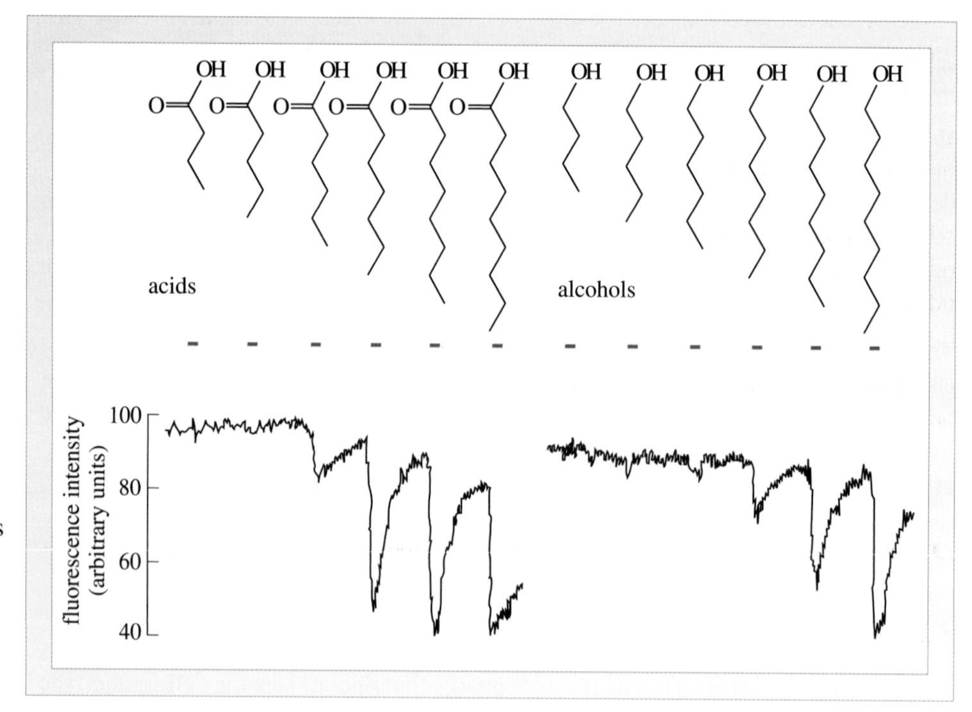

Figure 2.12 Fluorescence changes observed when an olfactory neuron loaded with fura-2 is exposed to various odorants for the same fixed periods of time (as shown by the length of the red bars).

❍ What do you notice about the fluorescence changes in Figure 2.12 in terms of both their responses to different odorants, and their time dependence?

● It appears that the size of the fluorescence change varies from molecule to molecule, with some molecules unable to effect any change whatsoever. For those odorants that effect a fluorescence change, response is time dependent; fluorescence diminishes a short time after exposure to the odorant, then gradually returns to the original level once exposure to the odorant is stopped.

2.7 Combinatorial coding of odours

As we noted earlier, it has been found that there are as many as 1000 genes for olfactory receptors in the mammalian genome. Given that we often think of our sense of smell as not being very highly developed, such a large number is remarkable because it means that the olfactory receptor genes comprise ≈ 1–3 per cent of the entire genome. Indeed, the olfactory receptor family is by far the largest mammalian gene family in the whole genome. For humans, however, about 60–70 per cent of these genes are pseudogenes – DNA sequences that are very similar to normal genes but which are not used to synthesise proteins. This compares to the relatively small 5 per cent content of pseudogenes in the rodent genome.

The general consensus is that humans can detect and discriminate between many thousands, indeed possibly an infinite number, of different odours. In 2001, by scanning the entire human genome, two groups – those of Doron Lancet in Israel and Sergey Zozulya in the USA – reported that there are approximately 350 full-length human olfactory receptor genes. Thus, humans have a repertoire of approximately 350 olfactory receptor proteins that are used to detect odorant molecules. Although this appears a large number, if a particular smell were triggered by the binding of odorant molecules to one specific receptor protein – a concept called the **labelled-line theory** (see Block 2) – then we would have a very restricted range of smell. Since each receptor responds to about three or four different odorant molecules, such a mechanism for odour coding would imply that our sense of smell would respond to about 1000–1500 different odorants.

Clearly, there has to be a different way, other than a labelled-line coding mechanism, in which the human brain uses the outputs from 350 olfactory receptors for odour detection and differentiation. Indeed, a major drawback of the one-receptor–one-odour mechanism is that a fixed number of receptors implies a fixed number of odours that we would respond to. How then could we sense the presence of a 'new' odour, one that had not been experienced previously?

The way in which the olfactory system copes with both these issues is via a **combinatorial odour code**. In this coding mechanism, the odour for any particular odorant molecule is represented by the activation of a combination of receptors, not just one receptor. It is the *pattern* of activated receptors that carries the information about the odour.

❍ What does this imply about the possible number of different receptors to which an odorant molecule will bind and activate?

● It means that any particular odorant could bind to and activate more than one different receptor.

This is, of course, something we noted earlier. The converse, we've already discovered, is also true – that each receptor has a range of molecules that it will accept.

To see how this combinatorial coding mechanism might work, take a look at Figure 2.13. This shows the response pattern from fourteen different mouse olfactory neurons, labelled S1, S3, etc., measured by the Ca^{2+} fluorescence technique, to several different odorants. The code for any one odorant is represented by the pattern of neurons that are activated together with the strength of the activation, here represented by depth of colour (the deeper the colour the stronger the activation). For example, nonanoic acid can activate olfactory neurons containing receptor S19 at very low concentrations but only activates neurons containing the S18 receptor at concentrations ten times greater.

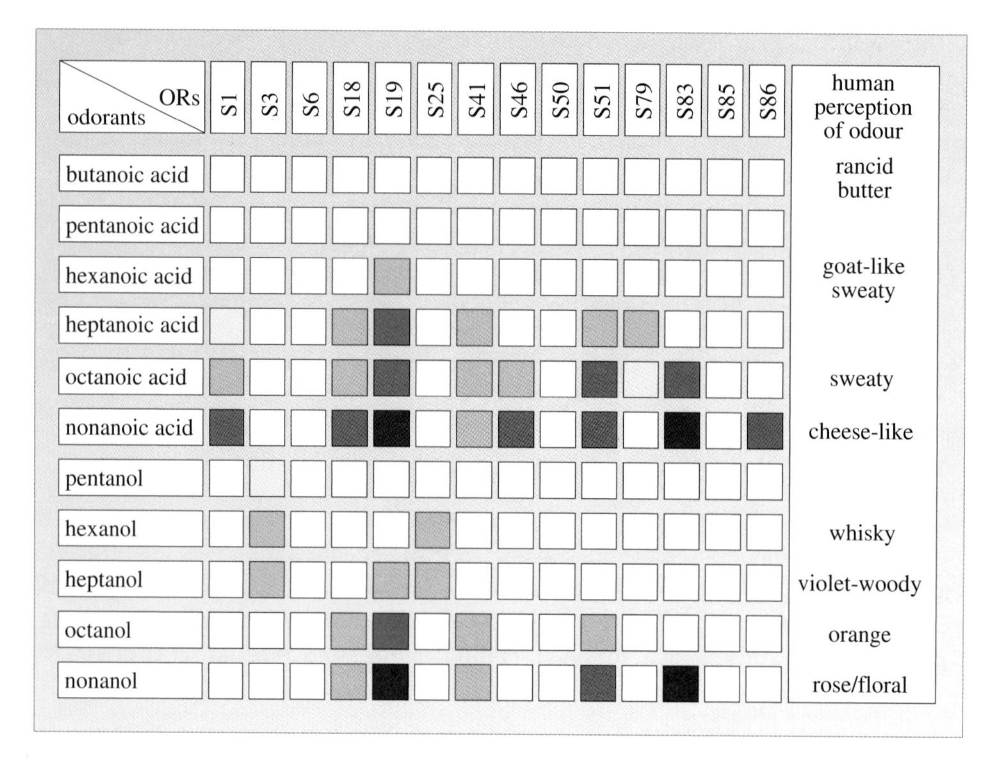

Figure 2.13 The combinatorial coding of odours. The strength of activation is represented by the depth of colour: pale yellow indicates the weakest and dark red the strongest. OR, olfactory receptor.

❍ Which of the two receptors is the more responsive to nonanoic acid, S18 or S19?

● As S19 is activated at lower concentrations it must be the more responsive.

So, in Figure 2.13, the neuron containing receptor S19 is represented by a deeper colour.

❍ Using colour strength as a cue, to which molecule does olfactory receptor S51 respond more strongly, octanol or nonanol?

● The depth of colour is stronger for nonanol, so S51 responds more strongly to this odorant.

What becomes clear from the odorant profiles shown in Figure 2.13 is that different odorants are recognized by different, and unique, combinations of activated olfactory receptors – and consequently olfactory neurons. No two odorants have identical profiles or receptor codes.

Another striking difference is that observed between the alcohols, which have floral fragrances, and the carboxylic acids, which tend to have rancid odours. From a molecular point of view, these two sets of molecules have similar structures. For those members of each class that contain equal numbers of carbon atoms, the only difference between the two is the group to which the —OH is attached: this is a CH_2 group in the alcohol and a C=O group in the corresponding acid (Figure 2.14). Yet, the two sets activate different combinations of receptors. For heptanol and heptanoic acid, for example, only one of the receptors was activated by both odorants, heptanol activating two further receptors and heptanoic acid a different five.

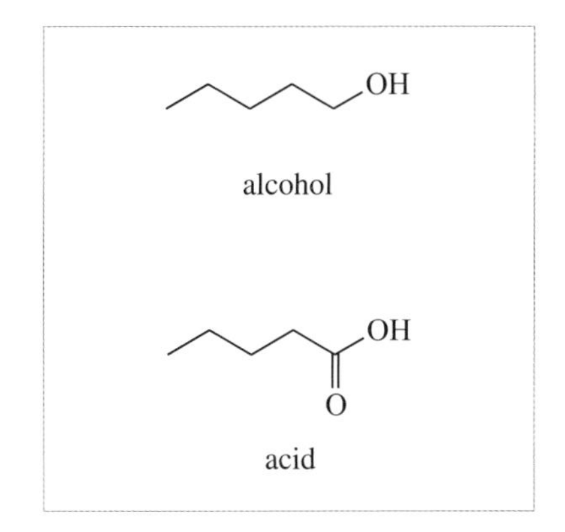

Figure 2.14 Comparison of the molecular structure of an alcohol with that of the corresponding acid.

In this way, by utilizing 350 receptors, the combinatorial mechanism of human olfactory coding can provide for an almost limitless number of molecules to be detected. Indeed, one estimate puts the number of odorants that we can discriminate at 10^5. To see how this may be so, let's examine a simple example.

❍ Assume that there are three different receptors, *a*, *b* and *c*. How many different codes involving all possible combinations of these three receptors are there? (Order does not matter here because the receptors are activated at the same time, so, for example, *ab* is the same as *ba*.)

● There are seven: the three different single codes – *a, b* and *c*; the three binary combinations – *ab*, *ac* and *bc*; and one ternary combination – *abc*.

So, for three receptors, while the labelled-line mechanism would code for only three odours the combinatorial mechanism has the potential to code for seven odours. We could, if we were so inclined, continue this idea to identify how many unique codes are possible with 350 receptors. However, it turns out that it is possible to express this mathematically using the following equation:

$$N = 2^n - 1$$

where N = number of codes and n = number of receptors.

❍ Use this equation to check the number of combinatorial codes from 3 receptors.

● $N = 2^n - 1 = 2^3 - 1 = 8 - 1 = 7$

The number of possible codes from 350 receptors can therefore be calculated to be $2^{350} - 1$, which can be expressed in terms of powers of ten to be a staggering 2.29×10^{105}! Such a number is many, many orders of magnitude more than the number of different odorant molecules the olfactory system will ever be called upon to detect. In fact it is more than the estimated number of particles there are in the entire universe (about 10^{80})!

We can also use the above equation to attempt an answer to the question, 'What is the minimum number of receptors needed to code for, say, a million (10^6) different molecules?' This can be expressed as follows:

$$2^n - 1 = 10^6$$

Since n must be an integer and 2^{20} = 1 048 576, then the minimum number of receptors is 20. This is clearly an order of magnitude less than the number of different olfactory receptors present in the human nasal epithelium, and it raises an important question, 'Given the energy demands associated with gene synthesis and protein expression, why has the olfactory machinery evolved with an apparently extraordinary level of redundancy within it?'

This is an issue of current debate. At the moment, we know very little about receptor expression patterns, especially about their expression across time. It is possible that our complement of olfactory receptors does not remain constant, some being important *in utero* while others may be important only in adolescence. This paints a dynamic picture of the olfactory system, something that cannot be deduced from the genome.

Alternatively, many of the olfactory receptor genes are very similar, containing only small differences between them, which means it isn't too difficult to come up with new receptors. So it may be that it is easier, more efficient even, to use a coding system based on lots of receptors than to develop a sophisticated one using fewer receptors. This would make olfaction similar to the immune system in which thousands of antibodies are made enabling the system to respond to antigens that have not been experienced previously.

Yet another possibility is that the role of many odorant receptors is to detect differences in concentration (we shall discuss the issue of concentration later). One problem that the olfactory system must address is the ability to recognize a particular odour over a wide range of intensity. As you will see, some odorants do change their quality with concentration, but for most we maintain a stable perception over a wide concentration range. Dedicating some receptors to responding to a specific concentration range, rather than to binding to specific types of molecule, is one way to overcome this problem.

A further possibility is that the number of receptors is related more to the ability to *discriminate* between closely related compounds rather than to the number of compounds we can smell. While the combinatorial coding method may allow for an essentially infinite number of smells to be *detected*, many of these codes are very similar, differing perhaps in just one or two receptors that are invoked. These odours may elicit essentially the same smell. Discrimination, rather than detection, is a distinction that is clearly important for a sensory system to make. The human olfactory system probably has the potential to detect a nearly limitless number of compounds (about 10 000 are currently known), but the number it can discriminate is not so large (about 3000 of the known 10 000). Current thinking is that most odorants activate about twenty or so receptors to give the distinctive perception of that odour – it is the pattern of activated receptors that enables discrimination.

We can get a 'feel' for the problem if we consider an example. Let's imagine that to discriminate between any two odours there must be a minimum of two receptors involved in the combinatorial code and that any discriminatively different code must contain at least two different activated receptors. How many different codes can be distinguished? Table 2.3 contains the relevant data for each situation up to seven receptors, *a*–*g*, involved in such coding.

❍ Given the above assumptions, how many receptors are needed to *detect* 15 odours? How many to *discriminate* between 15 odours?

● Only 4 receptors are needed for detection but 7 are needed for discrimination between 15 odours.

Quite clearly, to discriminate between odours requires more receptors than does detection. You can also begin to appreciate from this example that, in order to discriminate between a relatively small number of odours, there is a large redundancy in the codes. With the conditions that we have applied, for 7 receptors

Table 2.3 Combinatorial codes that differ by two receptors.

Number of available receptors	Total number of codes	Number of codes that differ by two receptors	Codes that differ by two receptors
2 (*a*,*b*)	3	1	*ab*
3 (*a*,*b*,*c*)	7	1	*ab*
4 (*a*,*b*,*c*,*d*)	15	3	*ab*, *cd*, *abcd*
5 (*a*,*b*,*c*,*d*,*e*)	31	5	*ab*, *cd*, *ace*, *bde*, *abcd*
6 (*a*,*b*,*c*,*d*,*e*,*f*)	63	11	*ab*, *cd*, *ef*, *ace*, *adf*, *bcf*, *bde*, *abcd*, *cdef*, *abef*, *abcdef*
7 (*a*,*b*,*c*,*d*,*e*,*f*,*g*)	127	15	*ab*, *cd*, *ef*, *ace*, *adf*, *bcf*, *bde*, *abcd*, *abef*, *acfg*, *adeg*, *bceg*, *bdfg*, *cdef*, *abcdefg*

almost 90% of the codes are of no use in discrimination. Nevertheless, the power of a combinatorial coding system is obvious, in that:

- it enables an almost infinitely large number of odours to be detected;
- it has the potential to detect a previously unknown odour;
- it has the ability to discriminate between odours.

Activity 2.2 Olfactory receptors

Now would be a good time to explore the *Olfactory receptors* activity on the CD-ROM. Further details are given in the Block 6 *Study File*.

2.8 Odour coding beyond the olfactory neuron

So far, we have confined our discussion to the coding pattern that is expressed at the level of the olfactory receptor and olfactory neuron. However, the axons of olfactory neurons converge onto roughly spherical features in the olfactory bulb, about 150 μm in diameter, called **glomeruli**. There are about 1000 glomeruli in each half of the olfactory bulb. It is in the individual glomerulus where axons from the olfactory neurons synapse with **mitral cells** (Figure 2.15 overleaf). Each mitral cell has a primary dendrite that extends into a glomerulus – consequently one mitral cell is connected directly to only one glomerulus – and an axon that, together with the axons from other mitral cells, forms the lateral olfactory nerve that extends into the **primary olfactory cortex**.

Our current understanding is that the axons of several thousand olfactory neurons converge on any one glomerulus, and that these neurons contain identical receptors and therefore have identical molecular receptive ranges. Within any one glomerulus, these axons form junctions with the primary dendrites of about 25 mitral cells.

❍ What would this suggest about the range of molecules to which a glomerulus should respond?

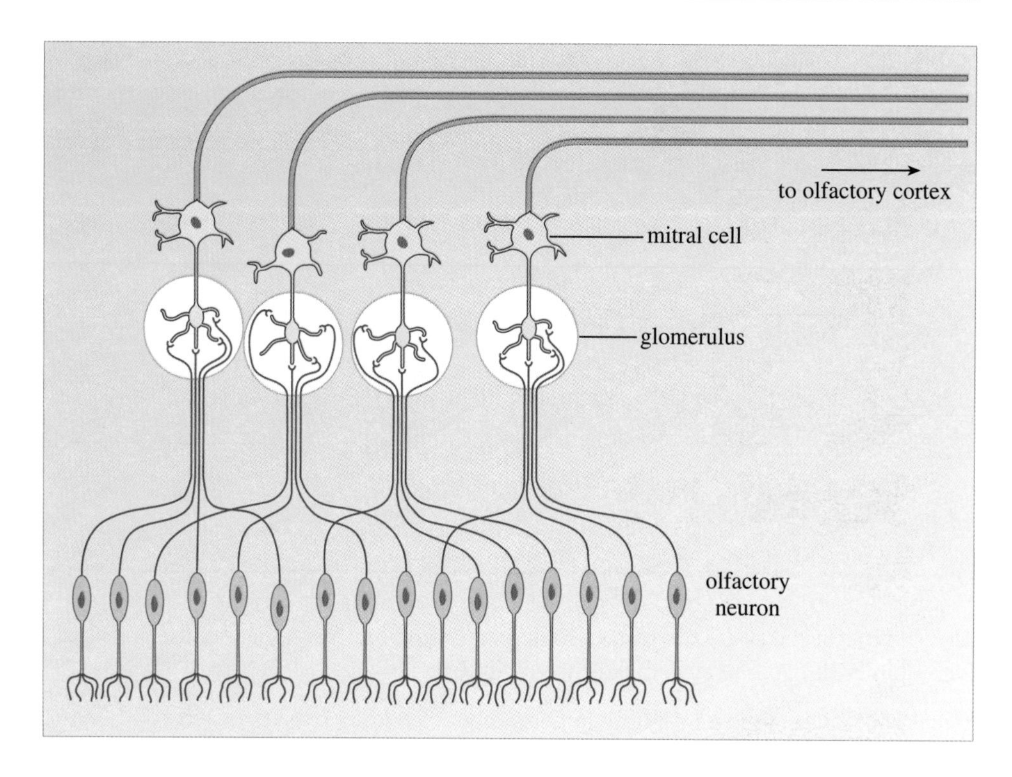

Figure 2.15 The glomerular connections between olfactory neurons and mitral cells.

- The glomerulus should be responsive to the same range of molecules that activates the olfactory neurons that converge onto it.

You might anticipate, then, that the glomeruli – and, by extension, the mitral cells – also combinatorially code for the odour of any particular odorant molecule.

❍ If this is correct, suggest how the glomerular activity in the olfactory bulb may differ when olfactory epithelium is exposed to two different odorants.

- If glomerular activity is combinatorially coded, each odorant should activate several different glomeruli and we would expect the two compounds to elicit two different patterns of glomerular activity.

Using a technique called **intrinsic signal imaging** this is precisely what has been observed. Intrinsic signal imaging relies on the detection of changes in the optical properties of tissue that are brought about by biochemical activity. For example, optical imaging of the olfactory bulb under red light (630 nm) produces signals that are due to changes in both blood volume and the oxygen saturation level of haemoglobin. Subtracting the data produced in the absence of any odorant from those of the same region when the olfactory epithelium is exposed to an odorant results in a picture of the upper 0.5 mm of the olfactory bulb, that is, the region corresponding to the glomerular activity (Figure 2.16).

The pictures in Figure 2.16, which correspond to about 5 per cent of the surface area of the olfactory bulb, reveal that different odorants effect different patterns of glomerular activity. For example, pentyl acetate, which has a fruity odour, activates a significant number of glomeruli in the imaged region whereas carvone, which has a caraway odour, only activates one distinct and different (white arrowhead) glomerulus. If you compare the patterns for octanal and pentyl acetate, molecules that are dissimilar both in terms of their structures and their odours, then you can see that both are able

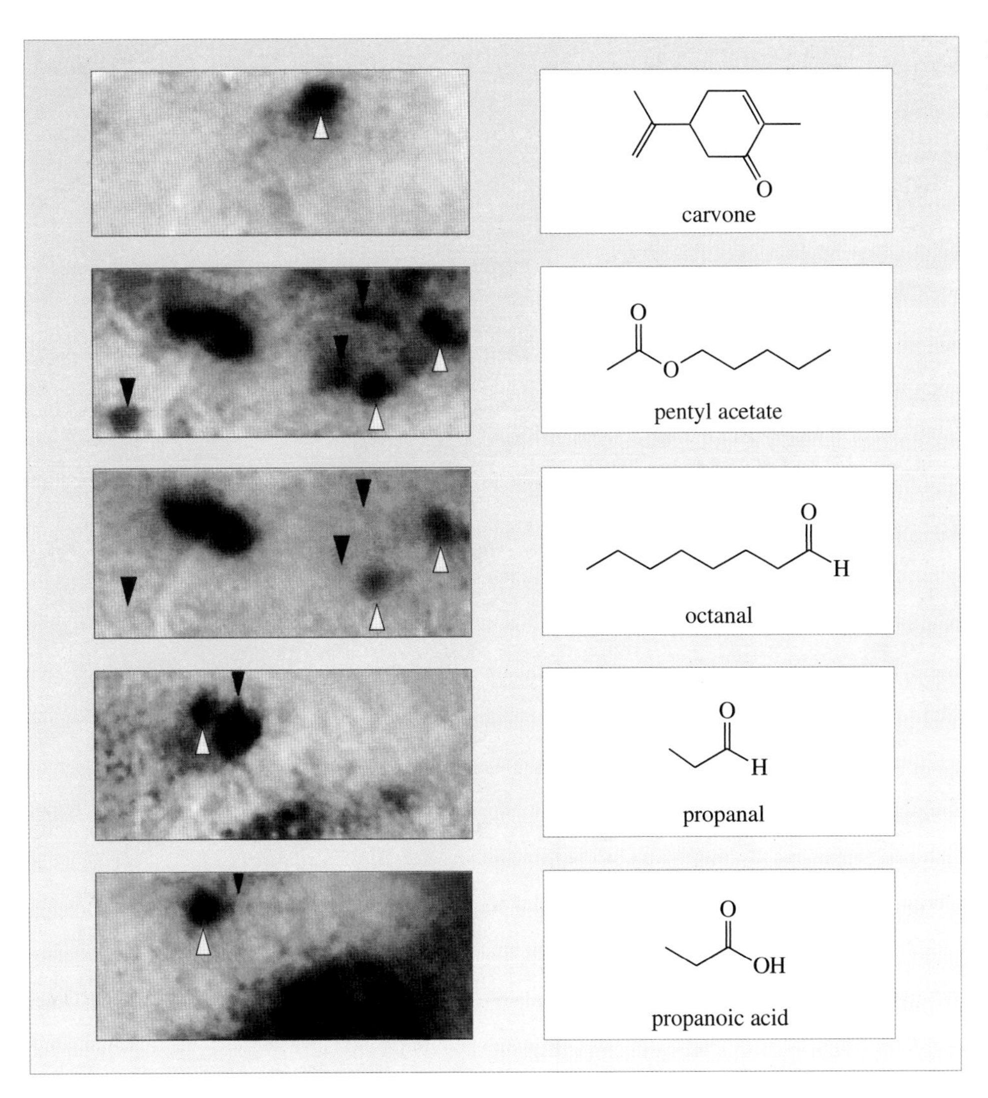

Figure 2.16 Glomerular activity of a section of the olfactory bulb in response to different odorants.

to activate common glomeruli (the broad dark patch in the upper left of each panel). However, it is also clear that there are glomeruli that are strongly activated by pentyl acetate but weakly or not at all by octanal (indicated by the arrowheads). It is also the case that odorants like propanal and propanoic acid, which have a common molecular feature (the carbon chain length), activate a common glomerulus (white arrowhead), though propanal also activates an additional glomerulus (black arrowhead).

Consequently, the representations of odours in the olfactory bulb are highly distributed across the glomeruli and we can see that:

- an individual odorant can activate many different glomeruli; and
- an individual glomerulus can be activated by many different odorants.

Of course, this is exactly the same observation that was made about the relationship between odorants and the olfactory receptors, so it would seem that the pattern of glomerular activation seen in the olfactory bulb also represents the combinatorial coding of an odour. That being so, glomeruli should display molecular receptive ranges, and that is precisely what has been observed. The responses of 40 glomeruli to a series of aldehydes whose carbon chain length varies from 3 to 10 (Figure 2.17 overleaf) show that each is tuned to at most just a few members of the series.

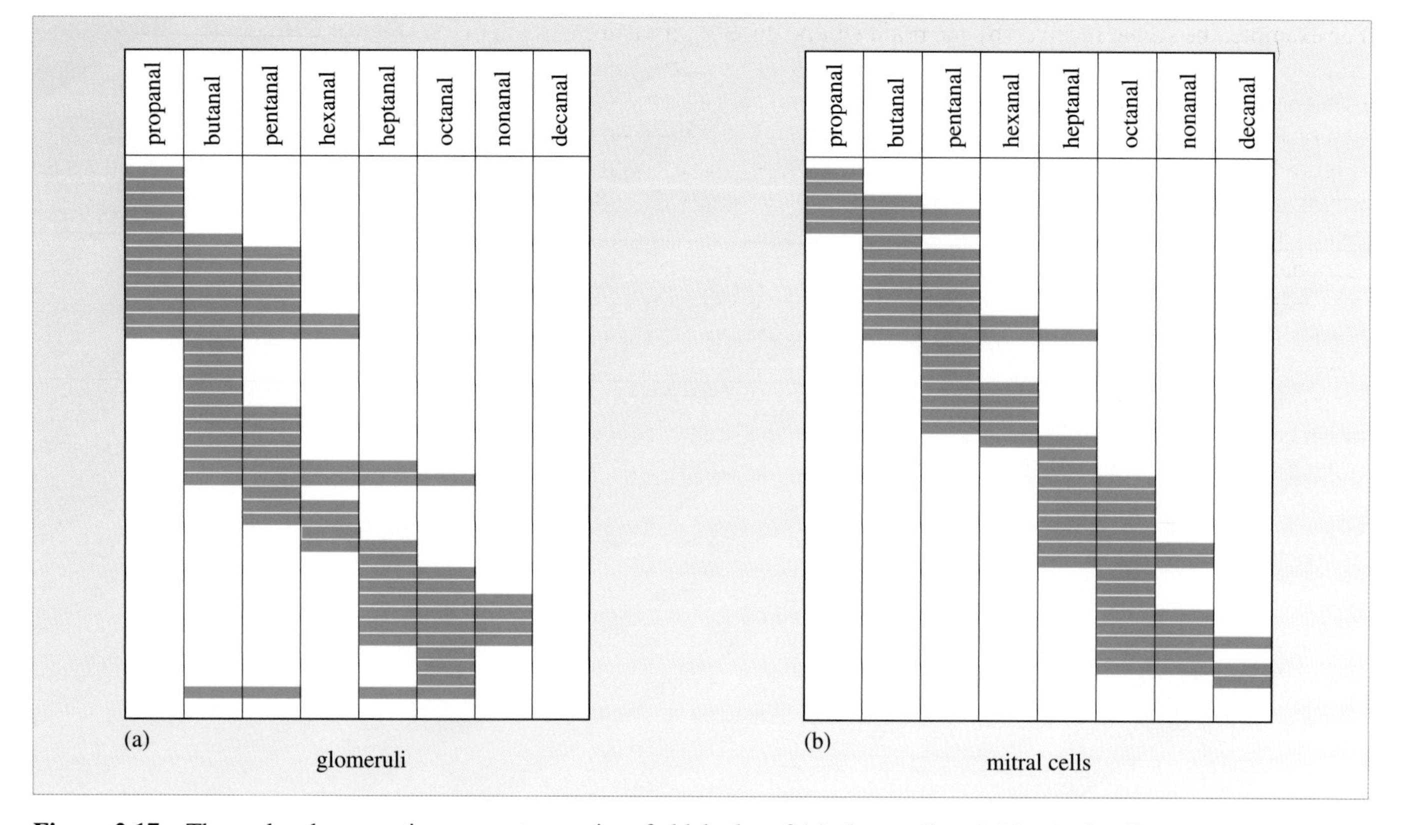

Figure 2.17 The molecular receptive ranges to a series of aldehydes of (a) glomeruli, and (b) mitral cells.

In fact, except for the last example and those activated by a single aldehyde, all the glomeruli responded to consecutive members of the aldehyde series.

Almost identical molecular receptive ranges are obtained when the mitral cell activity – brought about by odorant stimulation of the olfactory epithelium – is examined. Mitral cell activity can be monitored using microelectrodes that measure the action potential voltage changes with time. One such trace, for a mitral cell that is activated by the 4-carbon aldehyde butanal, is shown in Figure 2.18. Clearly, exposure to butanal brings about an almost instantaneous firing of the mitral cell. When the electrical response of several mitral cells to a range of aldehydes is measured in this way, the pattern of activity shown in Figure 2.17b emerges. This has a striking resemblance to that of the glomeruli shown in Figure 2.17a, and it demonstrates that the pattern of response observed at the level of the olfactory receptors is maintained at the level of electrical communication with the olfactory centres of the brain.

However, such communication with the higher olfactory centres is certainly more complicated than a mitral cell simply acting as a 'relay' for the signal produced by an olfactory neuron. First, the olfactory centres receive signals from many mitral cells. As with any ensemble, the output of the group contains additional information over and above the simple sum of the outputs of the individual mitral cells.

relative odorant concn

1

3

10

30

100

1 s

Figure 2.18 The voltage–time trace of a mitral cell responsive to butanal. The horizontal bar represents the period of exposure to butanal. The trace for respiration is shown below the activity trace. The arrows identify the beginning of inspiration immediately after exposure to butanal.

For example, the signal received by the brain will be different if the mitral cells fire randomly as opposed to being synchronized. Second, and related, is the 'cross-talk' that is possible between different glomeruli and mitral cells due to the presence of two types of interconnecting cells, **periglomerular cells** and **granule cells** (Figure 2.19). The periglomerular cells form synaptic interconnections, some of which are inhibitory and some excitatory, between the primary dendrites of mitral cells. Granule cells also form both excitatory and inhibitory synapses between the secondary dendrites of mitral cells and also with the axons of the mitral cells (which are the axons that form the olfactory tract). There are also synapses with efferent fibres. In this way, the mitral cell output is highly organized. Such organization has been observed in the locust (Figure 2.20). Initial exposure to an olfactory stimulus produces an initial intense burst of action potentials in the projection neuron (the insect equivalent of a mitral cell). Upon prolonged exposure to the odorant, the firing rate diminishes but the spike time precision increases markedly.

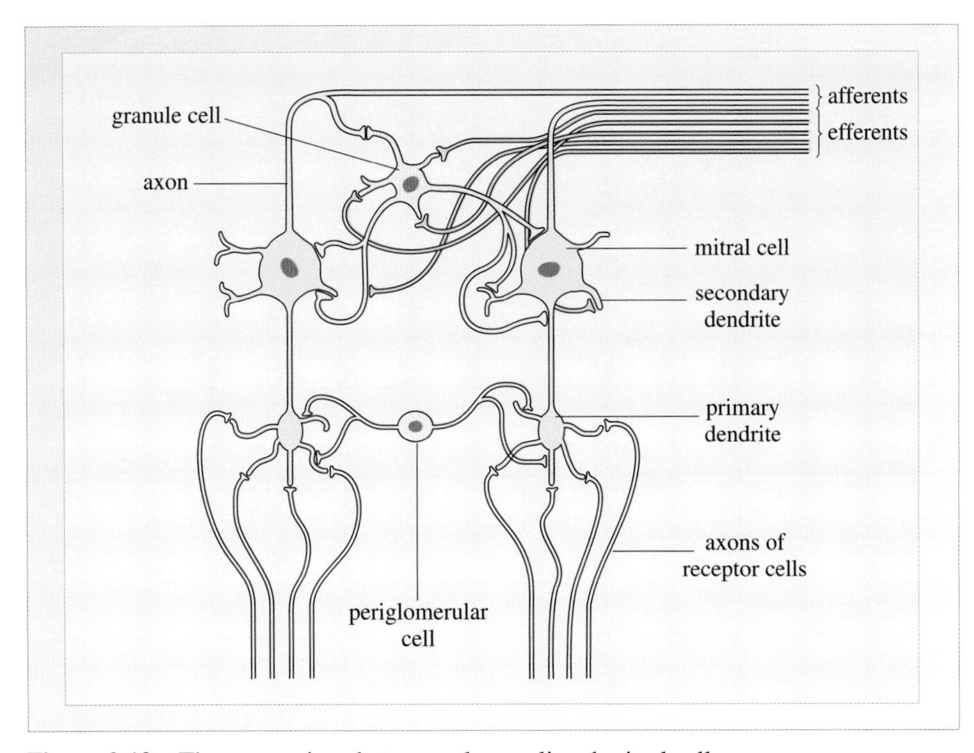

Figure 2.19 The connections between glomeruli and mitral cells.

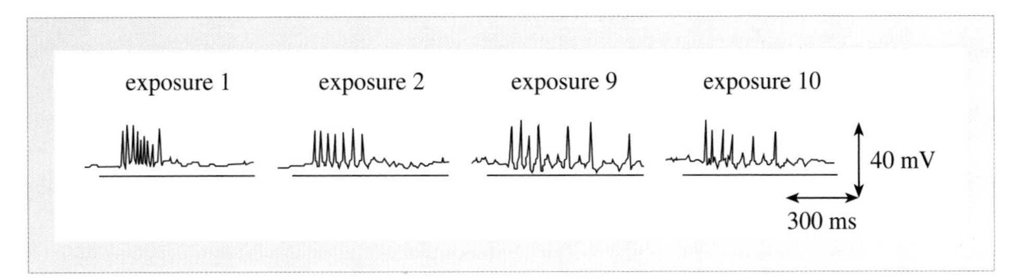

Figure 2.20 The change in locust neural output due to odorant exposure.

Currently, we still don't know how the brain decodes these patterns of electrical activity to represent them as a smell. As the Japanese scientist Kensaku Mori has written:

> When our knowledge of the olfactory cortex and higher olfactory centres advances, we might be able to determine why roses have a pleasant scent, whereas sweaty socks smell bad.

However, it seems probable that, alongside the combinatorial representation of odour, the way these patterns change with time also serves as a coding variable. Mori has proposed a possible mechanism by which the neurons in the olfactory cortex respond to the coordination (or lack of it) of the signals carried by the axons of the mitral cells. Imagine that the axons of two or more mitral cells from different glomeruli synapse onto the same neuron in the olfactory cortex (Figure 2.21). This cortical neuron can then function as a combination detector and its activity will represent the combined activity of two glomeruli. So, when an odour is detected, if the signal outputs of two mitral cells, say A and B, are synchronized then the probability that the cortical neuron that synapses with these will be activated is greatly increased.

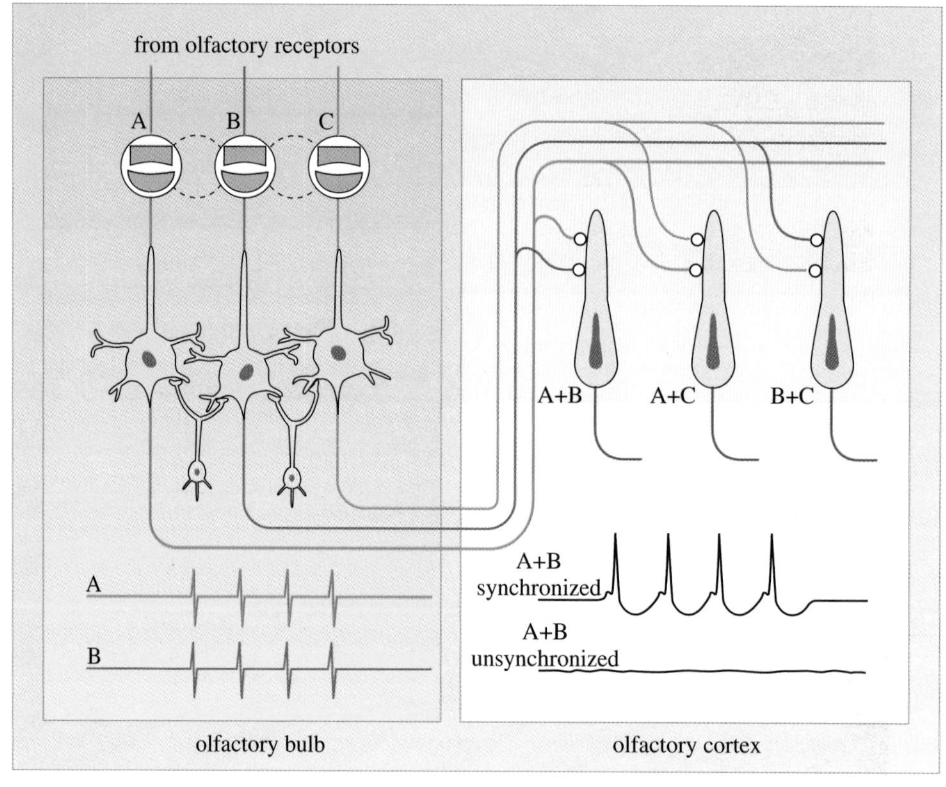

Figure 2.21 A possible mechanism for decoding olfactory signals in the olfactory cortex. When the signals of mitral cells A and B are synchronized, the olfactory cortical neuron can fire; when they are not synchronized, the olfactory cortical neuron does not fire.

Box 2.2 Smell and functional imaging

While much remains to be done to understand this higher order processing of the olfactory signals, functional imaging – using both PET and fMRI – is beginning to shed light on brain activity in response to odorant exposure. Figure 2.22 shows the regional distribution of brain activity in response to vanillin, a compound that has a vanilla odour. A more detailed examination of the activity in slice 5 with different odorants shows that the pattern of activation is odorant dependent (Figure 2.23).

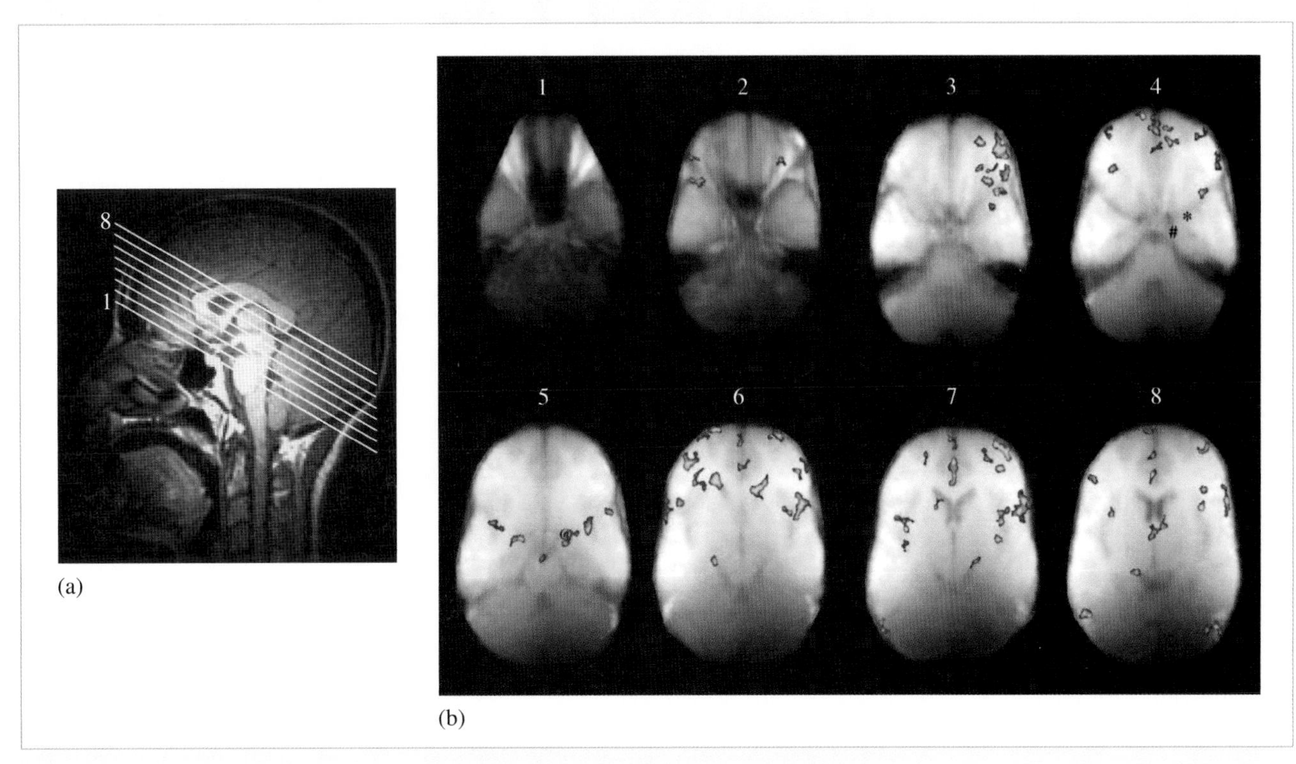

Figure 2.22 (a) A lateral view showing the positions of the slices in the fMRI. (b) The regional distribution of brain activity as visualized by fMRI due to exposure to vanillin.

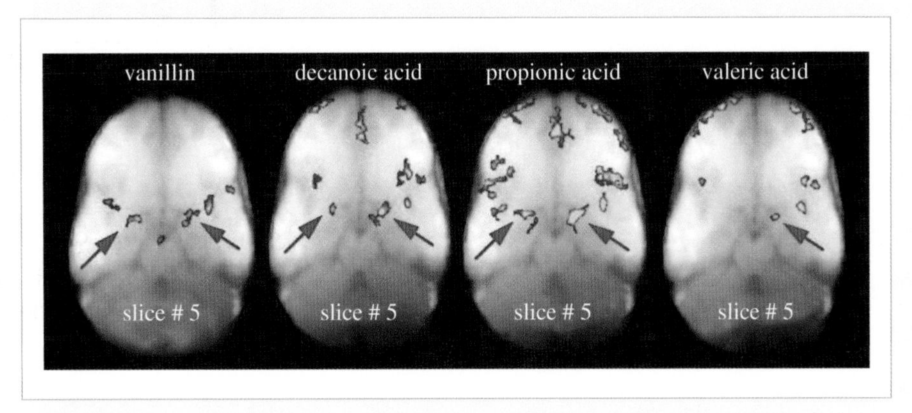

Figure 2.23 Variations in brain activity due to different odorants. (Propionic is another name for propanoic and valeric is another name for pentanoic.)

When subjects were asked to name common odours, a PET study revealed activity in five brain areas (Figure 2.24): the left cuneus (a), the right anterior cingulate gyrus (b), the left insula (c), the left anterolateral cerebellum (d) and the right postmedial cerebellum (e), but only the activity in the left cuneus was found to be specific to odorant identification.

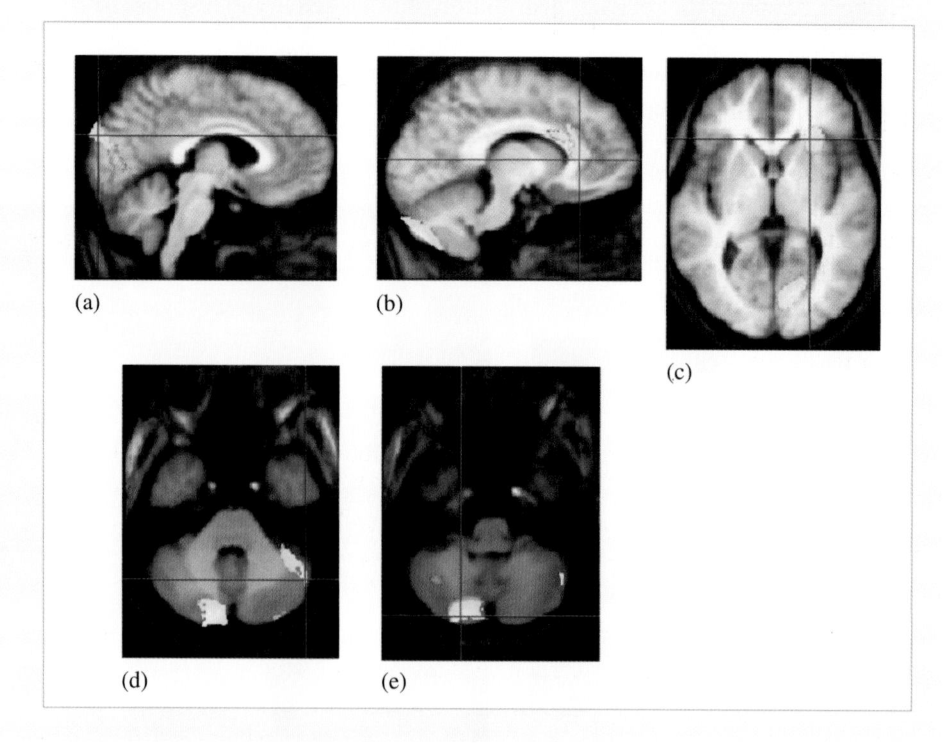

Figure 2.24 A PET image showing the regions of brain activated when odorants are identifed.

2.9 Summary of Sections 2.6–2.8

The binding of an odorant to an olfactory receptor initiates a cascade of molecular events that brings about an influx of Ca^{2+} and Na^{+} ions into the olfactory neuron. The membrane potential of the neuron changes from –65 mV to –45 mV, a depolarization which evokes action potentials. The pattern of olfactory neurons that are activated provides a combinatorial code for the odour. Since we have about 350 different olfactory receptors, the number of different molecules that could be detected by combinatorial coding is almost infinitely large. This coding mechanism is powerful because it also allows for the discrimination between a large number of different odours.

The axons of olfactory neurons terminate in synapses to mitral cells in the glomeruli of the olfactory bulb. Each glomerulus receives axons from several thousand of the same type of olfactory neuron, and each glomerulus is associated with about 25 mitral cells. The combinatorial coding mechanism is evident at the level both of the glomerulus and of the mitral cell. The axons of the mitral cells form the lateral olfactory tract that leads to the primary olfactory cortex. The signals carried by these axons are highly coordinated by the action of periglomerular and granule cells. It is thought that the olfactory cortical neurons act as decoders of synchronized inputs from the mitral cells. Although it is currently unclear how the brain decodes the olfactory inputs, fMRI and PET are able to image the brain activity brought about by odour stimulation.

2.10 The effect of concentration on odour perception

Although the perception of many odours remains constant with changes in the concentration of the odorant molecule, some odorants have pronounced differences in the smell perceived at different concentrations. A well-known example of this is indole, which has a pleasant jasmine-like smell at low concentration but an unpleasant faecal stench at higher concentrations. Can the description of smell in terms of a combinatorial representation of odorant receptor interactions account for such an observation?

To begin to see how this might be possible, you need to know that the interactions between any two molecules, A and B, depend on several factors, including the number and strength of the bonding interactions (hydrogen bonding, London forces, etc.) as well as the concentration of the two molecules involved. The interaction is written as:

$$A + B \rightleftharpoons A{\bullet}B$$

where A•B is the complex formed from the relatively weak association of A and B, and the $\rightleftharpoons$ symbol means the process is reversible, that is, A and B can associate to form A•B, and A•B can dissociate into A and B. The process is *dynamic*, with molecules of A and B continuously binding to form A•B, while an A•B complex dissociates to form separate molecules of A and B. If the concentration of *either* A or B increases then more A•B is formed.

The effect of concentration of the odorant molecule on the response of the olfactory receptor has been studied *in vitro*. The genes for several different olfactory receptors can be inserted, separately, into human embryonic kidney cells, that is, into cells that do not normally respond to odorant molecules. The effect of odorants on these cells was then examined using the fluorescence method that responds to changes in the concentration of Ca^{2+}; the greater the change in the cellular Ca^{2+} concentration the greater the change in fluorescence.

❍ From what you know about the mechanism of olfaction, how does the cellular Ca^{2+} concentration change when an odorant binds to its receptor?

● Odorant binding results in the opening of a Ca^{2+}/Na^{+} ion channel and an influx of Ca^{2+} into the cell.

Figure 2.25 (overleaf) shows how the fluorescence changes when the cells containing one such receptor were exposed to vanillin.

❍ From Figure 2.25a, what can you deduce about the Ca^{2+} flux upon exposure to vanillin?

● The Ca^{2+} flux is (a) transient and (b) dependent upon the concentration of vanillin.

❍ Using the data in Figure 2.25b what additional observation can be made about the Ca^{2+} flux?

● The Ca^{2+} changes reach a maximal value; increasing the vanillin concentration beyond 300 μM brings no further increase in Ca^{2+} flux.

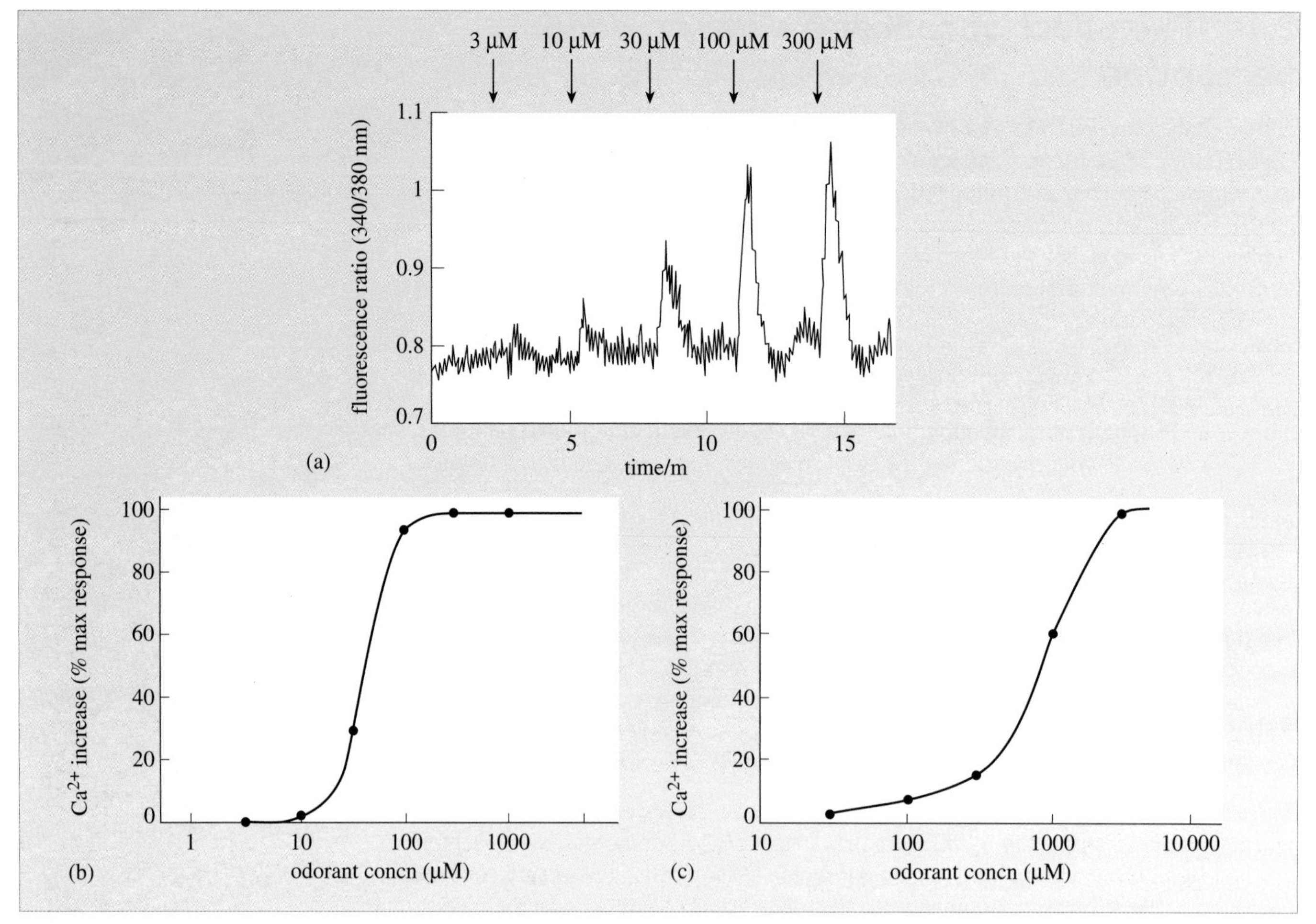

Figure 2.25 (a) Fluorescence changes observed when embryonic kidney cells that express an olfactory receptor are exposed for 20 s to various concentrations of vanillin (μM is micromolar or 10^{-6} molar; m = minutes). (b) Graphical representation of the data in (a) expressed as a percentage of the maximal effect (note that the trace for the 1000 μM concentration is not shown in (a); also that the concentrations (concn) are represented by a logarithmic scale). (c) Graphical representation of the effect of vanillin on cells that contain a different receptor.

The receptor is said to saturate, and **receptor saturation** is a general property of molecular interactions between small molecules, such as odorants, and fixed quantities of receptor proteins. A useful measure of the affinity between an odorant molecule and an olfactory receptor is the odorant concentration that brings about a change in effect that is half the maximal effect observed. This concentration is called the $\mathbf{EC_{50}}$. (The EC_{50} is a general measure of the affinity between any molecule and any receptor.)

❍ When comparing two EC_{50} values, which represents the stronger receptor binding affinity, the lower EC_{50} or the higher?

● The lower EC_{50} represents stronger receptor binding affinity because it indicates that a smaller concentration is able to bring about the same response as a compound that requires a larger concentration.

❍ Use the two graphical representations in Figure 2.25b and c to determine the EC_{50} values of vanillin for the two receptors.

● The EC_{50} values are approximately 40 μM and 900 μM.

It is obvious from these two values that vanillin binds to one type of receptor about twenty-five times more strongly than it does to a different type of receptor. Because of this, it is able to elicit a biological response from the first receptor type at concentrations much lower than it does from the second type of receptor. This is what we meant earlier when we said that some of the receptors might be concentration detectors. Moreover, at a concentration of, say, 1 mM (1000 μM) we can see from Figure 2.25b and c that the first receptor type is activated maximally whereas the second receptor type is activated at just over half the maximal response. Even with just these two receptors as our illustration, we can see that exposure to low odorant concentrations will activate only those receptors to which the odorant binds strongly. At higher odorant concentrations, additional receptors to which the odorant has weaker affinity are then brought into play. Consequently, the combinatorial pattern of receptor activity is able to change with odorant concentration and it is entirely possible for the perceived odour to change as well.

This effect can also be seen at the level of the glomeruli. Take a look at Figure 2.26, which shows the glomeruli in one particular area of the olfactory bulb that are activated by differing concentrations of pentyl acetate (a compound that has a distinct pear-like aroma).

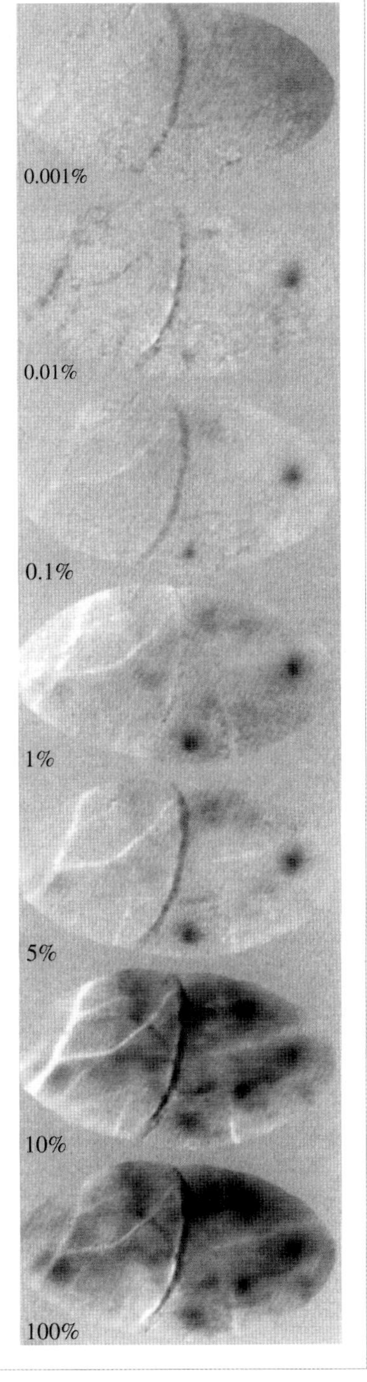

Figure 2.26 Glomerular activity in response to varying concentrations of pentyl acetate.

❍ What do you notice about the pattern of glomerular activity with changing concentration?

● At low concentration only one glomerulus is activated. With increasing concentration, additional glomeruli are involved and a much broader pattern of activity is observed.

This is entirely explicable from the odorant molecule/olfactory receptor model. At low concentrations, only those receptors to which pentyl acetate binds strongly are activated; these in turn activate their associated glomerulus. At higher concentrations, those receptors that have weaker affinity for pentyl acetate are activated, and at the highest concentrations, receptors of lowest affinity are activated.

So, as the concentration of pentyl acetate varies, its glomerular coding pattern changes. While for pentyl acetate this does not elicit a change in the perceived odour, for other substances it is entirely likely that it could. However, our current knowledge is insufficient to account for some odorants being perceived differently as their concentration changes while for others there is simply a change in odour intensity.

2.11 Adaptation

No doubt, we are all well aware of what happens when we are exposed to an environment containing a novel smell. Initially, our ability to sense the odour is heightened, but quite rapidly we become accustomed to the smell and our ability to sense its presence is diminished. This is not because the smell disappeared, as you will have noticed from the response of someone who subsequently enters the room. Rather it is because our olfactory system has adapted to the presence of the odour and no longer responds to it in the same way. Sensory adaptation was first introduced in Blocks 1 and 2.

How does this happen? There appears to be several mechanisms. The first, described as **short-term adaptation**, is the type of adaptation observed when olfactory receptor neurons are exposed to very short pulses of an odorant molecule.

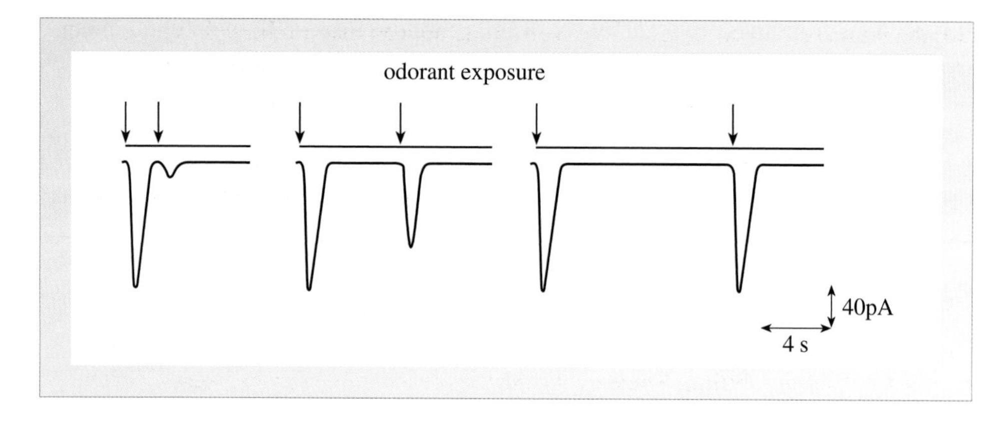

Figure 2.27 Short-term adaptation exhibited by the membrane currents of an olfactory neuron that are observed upon 100 ms exposure to the odorant cineole.

The effect is illustrated in Figure 2.27. Here, an olfactory neuron has been exposed to two 100 ms pulses of cineole (the major component of eucalyptus oil) that are spaced varying lengths of time apart.

❍ What do you notice about the response of the olfactory neuron to the second pulse of cineole?

● The closer the second cineole pulse is to the first, the smaller the membrane current. When the second pulse of cineole is separated from the first by about 10 s, the membrane responses of the two pulses are identical.

The half-life for recovery from short-term adaptation is about 4–5 s, and this type of adaptation has been shown to be related to the flux of Ca^{2+} ions across the olfactory receptor neuron. In fact, when the time-course for the recovery of the neuron from the Ca^{2+} influx was measured using a fluorescence technique and compared with the time-course for recovery from adaptation (Figure 2.28), the two were found to be essentially identical.

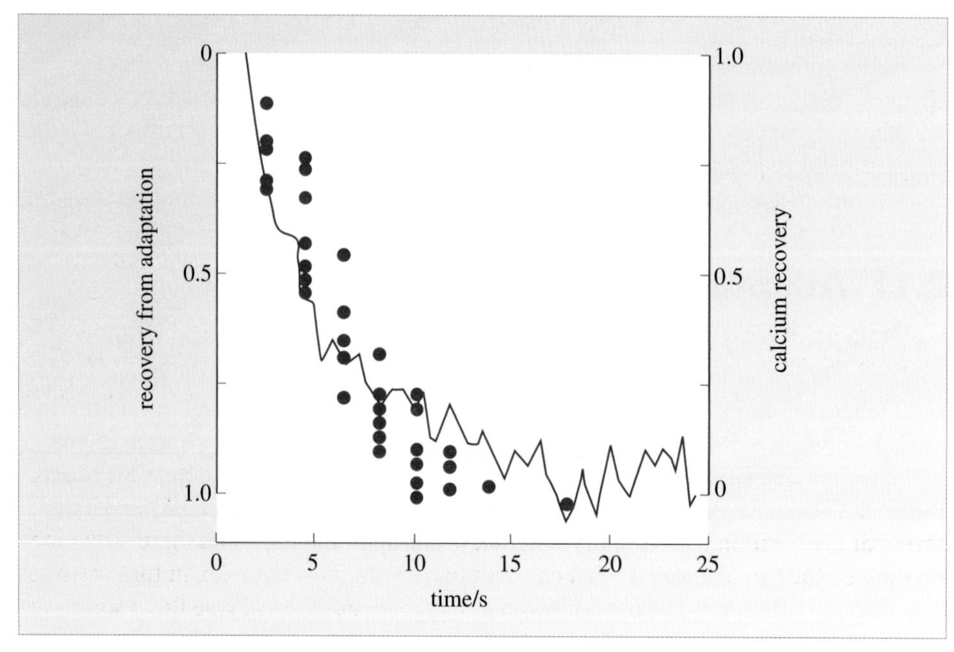

Figure 2.28 Time courses for Ca^{2+} recovery (solid line) and recovery from short-term adaptation (red dots) of an olfactory receptor neuron upon exposure to cineole.

So, the increased levels of Ca^{2+} ions within the neuron that are brought about upon short, pulsed exposure to the odorant appear to be responsible for the short-term adaptive response. This can be confirmed by observing the response of an olfactory neuron into which a Ca^{2+} chelating agent has been inserted (the chelating agent binds so strongly to Ca^{2+} that it effectively removes the Ca^{2+} ions from any of the biochemical processes within the cell). This is shown in Figure 2.29. In the presence of the Ca^{2+} chelator there is no adaptive response.

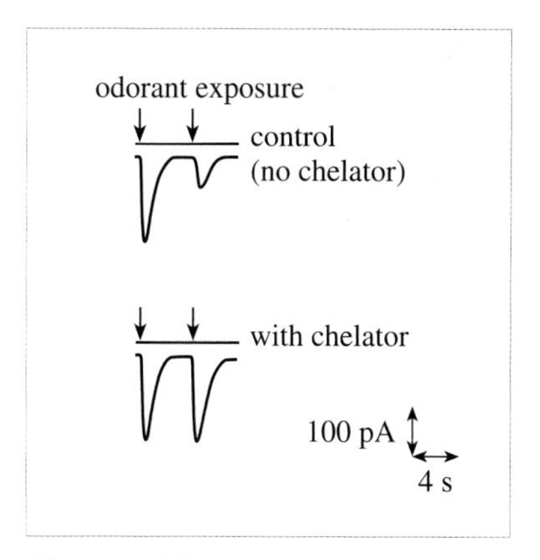

Figure 2.29 Response of an olfactory neuron to cineole in the absence and presence of a Ca^{2+} chelator.

How do the increased levels of Ca^{2+} ions within the cell result in short-term adaptation? It is thought that the Ca^{2+} ions bind to an as yet unidentified protein, thereby reducing the affinity of cyclic AMP for the Ca^{2+} ion channel, a process called feedback inhibition. Since cyclic AMP is responsible for the opening of the channel, reduced cyclic AMP affinity results in reduced channel opening, and in turn reduced neuronal firing. Only when the transient increase in Ca^{2+} is cleared from the cell can cyclic AMP function efficiently and the response to the odorant return to normal (Figure 2.30).

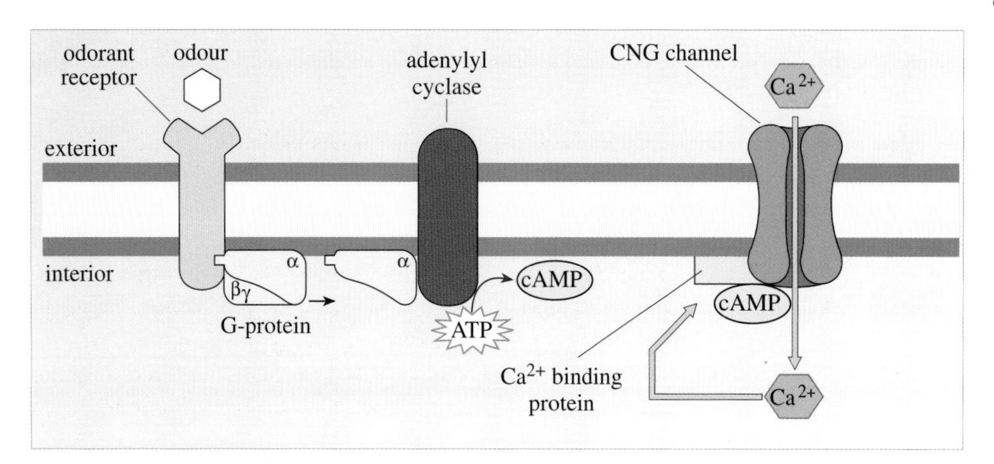

Figure 2.30 How Ca^{2+} levels are responsible for short-term adaptation. (CNG = cyclic nucleotide gated.)

A second type of adaptive response is called **odour response desensitization**. This is the response observed upon longer periods of exposure to an odorant, and is illustrated by Figure 2.31 (overleaf).

It is clear from this figure that initial exposure to cineole produces a sensory response that declines even in the continued presence of the odorant. Moreover, following a 10 s interval after removal of the initial cineole stimulus, a second exposure to the odorant produced a strongly attenuated response. The time-scale involved in desensitization is clearly longer than that for short-term adaptation. In fact, it requires about 1–1.5 minutes for the olfactory sensory neuron to fully recover from the desensitization brought about by an 8 s exposure to cineole.

What is the mechanism responsible for this kind of desensitization? It would appear to be another, distinct, Ca^{2+} feedback pathway (Figure 2.32 overleaf). The elevated levels of Ca^{2+} within the olfactory cell that occur upon odorant binding result in the binding of Ca^{2+} to a protein called calmodulin (CaM). This complex in turn activates a protein, CaMKII kinase, that inhibits adenylyl cyclase, the enzyme that forms cyclic AMP from ATP. Since cyclic AMP production is decreased, the Ca^{2+} ion channel that is activated by cyclic AMP is inhibited and Ca^{2+} ion influx is diminished.

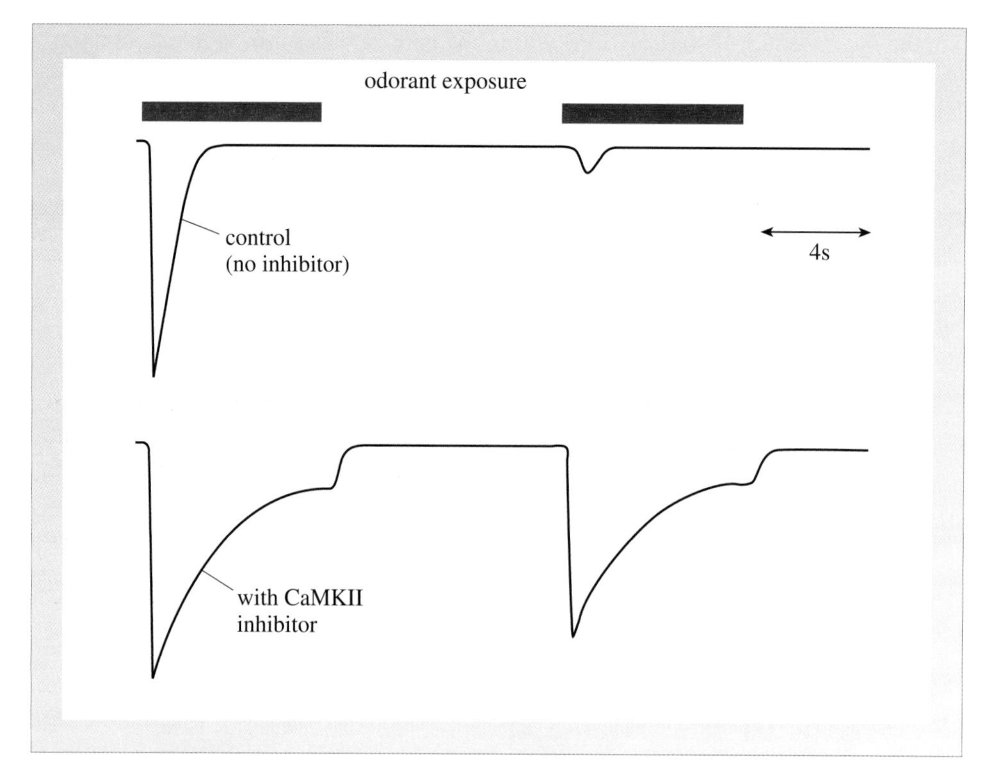

Figure 2.31 Adaptive response of an olfactory receptor neuron to prolonged (8 s) exposure to cineole in the absence and presence of a CaMKII inhibitor.

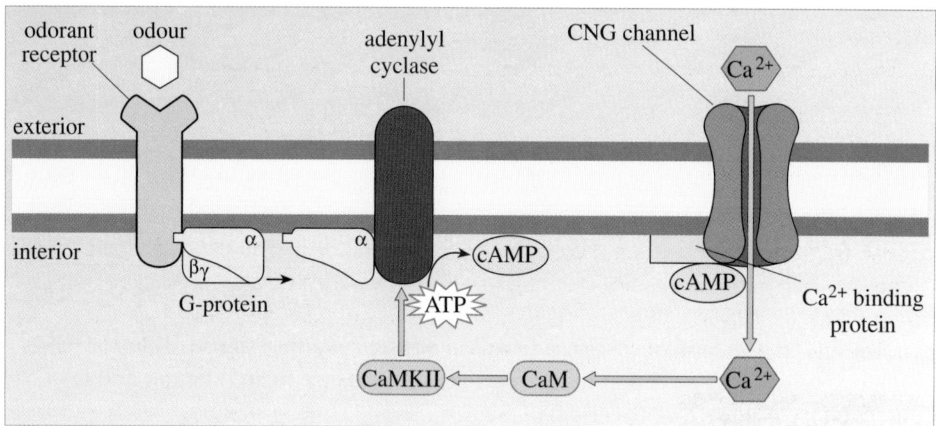

Figure 2.32 Proposed mechanism of odour response desensitization.

When the CaMKII kinase was selectively inhibited, so that the formation of cyclic AMP could function normally, the rate of desensitization was significantly reduced (Figure 2.31) and the response to a second cineole exposure was almost totally restored to initial exposure levels.

A third type of adaptation is known, at least for the salamander, and this response is called **long-lasting adaptation**. In this case, even following relatively short exposures (100 ms) to odorant some olfactory receptor cells exhibit an adaptive response that lasts on the minute timescale (Figure 2.33), and normal cell response is only recovered after six minutes in an odour-free environment. The molecular basis of this type of adaptation is presently unclear, although it is believed to involve Ca^{2+} ion fluxes brought about by cyclic GMP (rather than cyclic AMP).

Why does our olfactory system respond adaptively in these ways? Presumably, under evolutionary pressure our sense of smell has become tuned to *changes* in our

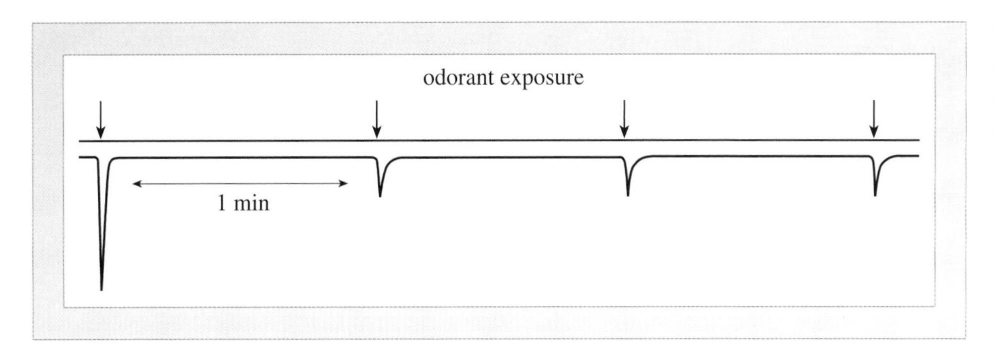

Figure 2.33 An olfactory neuron that displays long-lasting adaptation upon exposure to odorant.

odour environment as a means of alerting us to opportunities and danger. If we were to respond without adaptation to every odour we encountered, then our brains would be subjected to an overload of information. Moreover, the onset of odour adaptation, and recovery from it, are both time-dependent and consequently they have critical roles in the temporal response of the olfactory system. Indeed, the time-dependence itself carries crucial information about the odour that can be used, for example, to help locate its source. This temporal information is therefore probably a part of the chemosensory code.

2.12 Odour discrimination

The mammalian olfactory system is remarkable in its capacity to perceive and discriminate between thousands of different odorant molecules and thousands of different odours. In fact, we are seldom exposed to single odorant molecules. Even when we enjoy the scent of a flower, jasmine (Figure 2.34) for example, we are sampling a diverse range of some two hundred odorant molecules, a selection of which are shown in Figure 2.35 (overleaf).

Many of these are not unique to the scent of jasmine; geraniol is also a component of rose fragrance, indole is present in orange blossom, and benzyl acetate – the main component of jasmine – is also found in hyacinth and gardenia. However, certain components – jasmone, methyl jasmonate and jasmine lactone – are unique to jasmine and, not surprisingly, these compounds have odours characteristic of jasmine.

Figure 2.34 *Jasmine officinalis.*

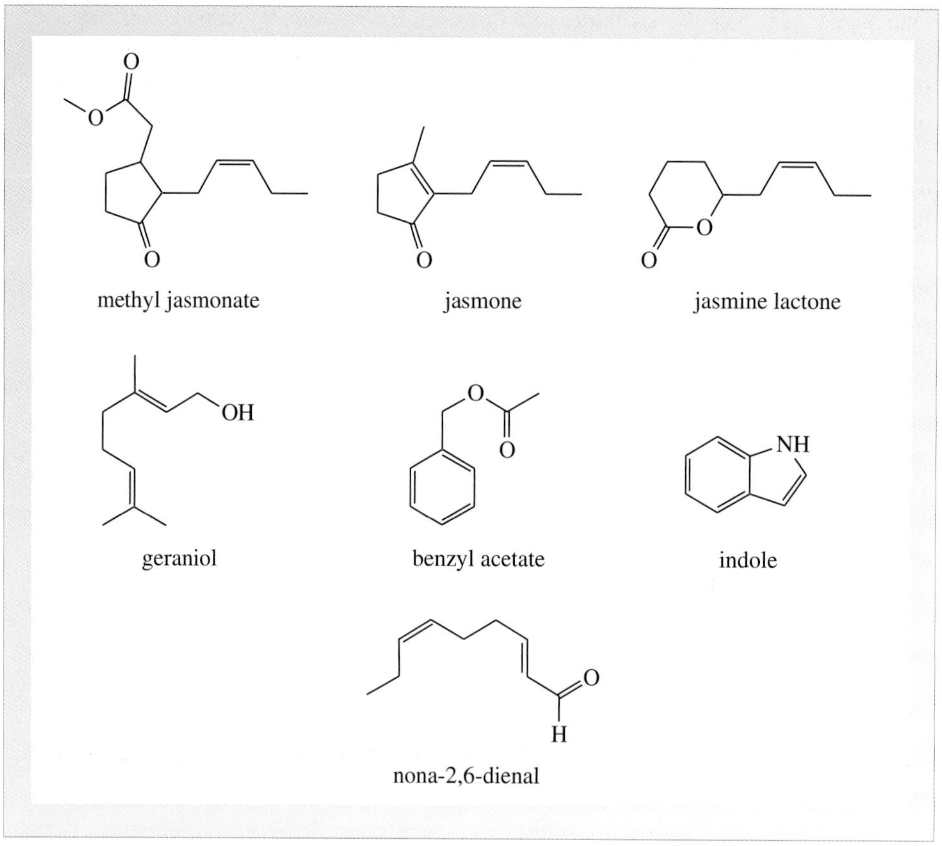

methyl jasmonate

jasmone

jasmine lactone

geraniol

benzyl acetate

indole

nona-2,6-dienal

Figure 2.35 Some of the odorant molecules found in *Jasmine officinalis*.

This raises an important question, namely, 'How well do humans distinguish between individual components in complex odour mixtures?' Attempts to obtain answers to this question have largely followed two lines of approach: (1) the ability to discriminate in mixtures odorants that have largely similar molecular structures, and (2) the ability to discriminate between mixtures involving structurally diverse odorants that have either similar or dissimilar odour qualities. Of course, any answer to this question has a bearing on our understanding of how we perceive the odour of a complex mixture.

First, then, how well do we discriminate between the odours of structurally similar molecules? This has been investigated using a method called the **forced-choice triangular test**. In this experimental paradigm, human subjects are presented with three samples, each of which contains an olfactory stimulus (which, depending on the experiment, may be a single odorant or a mixture of odorants) dissolved in an inert (so far as odour is concerned) solvent. Two of the samples contain identical stimuli, the third contains a different stimulus. By comparing the odours of the three solutions, the subjects are asked to identify which bottle contains the 'odd' stimulus. To minimize error, the experiment is repeated on four or five occasions at intervals of one to three days apart.

❍ Suggest a reason for leaving an interval of a day or so between repeat experiments.

● Such an interval should remove the possibility of adaptation affecting the ability to discriminate between the odours.

One aspect of molecular structure that has been examined in this way is the size of the molecule and, in particular, the length of the carbon chain (see Figure 2.36). For each class of odorant, the participants were presented with stimuli that consisted of two odorants that differed by one or more CH_2 groups (see Figure 2.37). A general finding of this approach is that humans find it much more difficult to discriminate between odorants that differ in size by just one CH_2 group than between those that differ by more than one CH_2 group. Two typical plots for ketones and alcohols, showing how the error in selecting the 'odd' stimulus varies with molecular chain length (n in Figure 2.36), are shown in Figure 2.38 (overleaf). It is obvious from these that humans have the ability to easily discriminate odorants that differ by three or more CH_2 groups, but that we are much poorer at discriminating between more closely related structures. An important point to note here is that such discrimination is not language related; that is, it is not dependent on the ability to *describe* the difference between the odours only to recognize that there is one. For example, in the acid series, acetic acid ($n = 0$) smells of vinegar, butanoic acid ($n = 2$) is commonly described as having a rancid butter aroma, and hexanoic acid ($n = 4$) has the odour of goats. However, to describe the differences between the odours of these acids in such terms demands an appropriate vocabulary and an ability to apply it to subtle differences in odour. The forced-choice triangular test circumvents this problem.

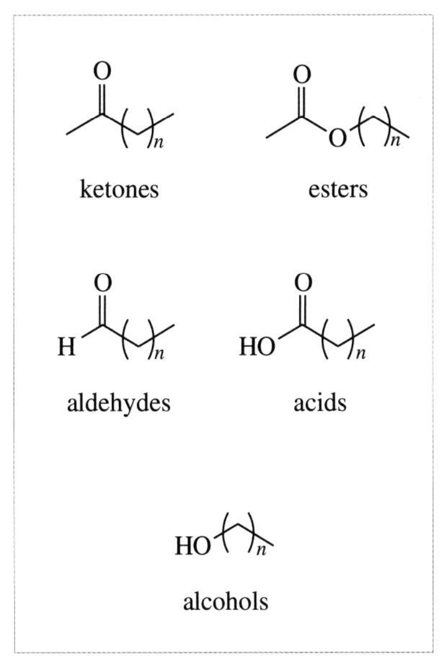

Figure 2.36 General structure of several classes of odorants investigated by the forced-choice triangular test (n here represents any number of intervening CH_2 groups).

❍ In terms of olfactory receptor theory, why might odorants of similar structure be more difficult to discriminate at the perception level?

● Odorants of similar structure should bind to a similar range of receptors. This would result in patterns of neural activity that are likely to be similar.

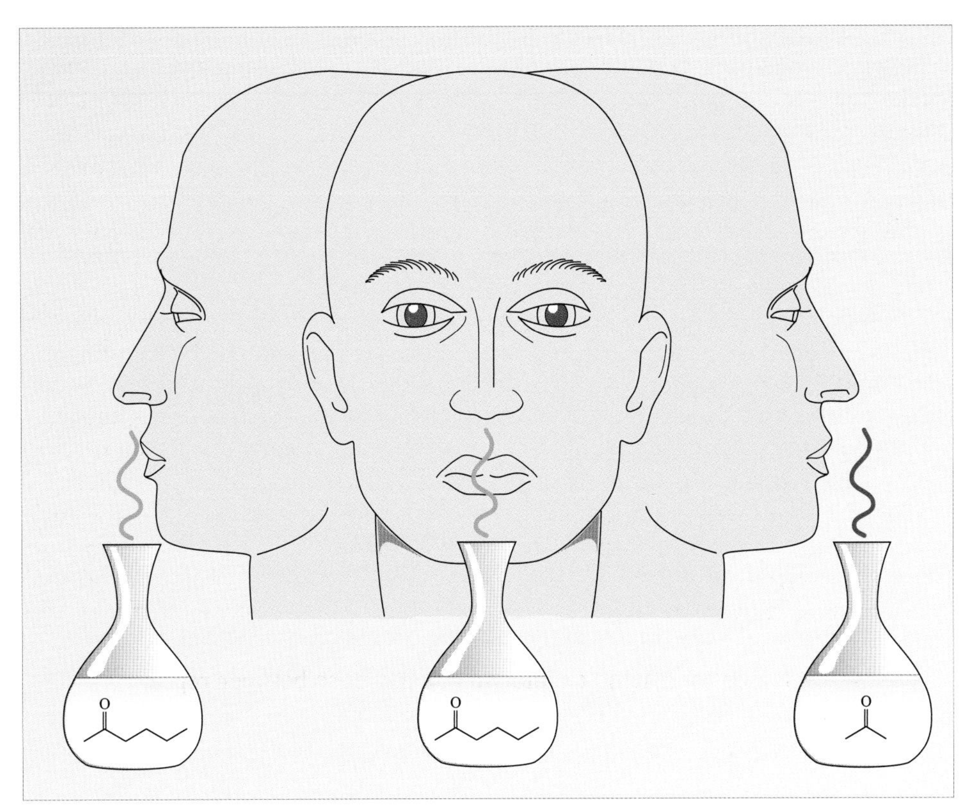

Figure 2.37 Schematic illustration of a forced-choice triangular test to discriminate between two ketones, $CH_3CO(CH_2)_nCH_3$, with $n = 0$ and $n = 3$, and which therefore differ by three CH_2 groups.

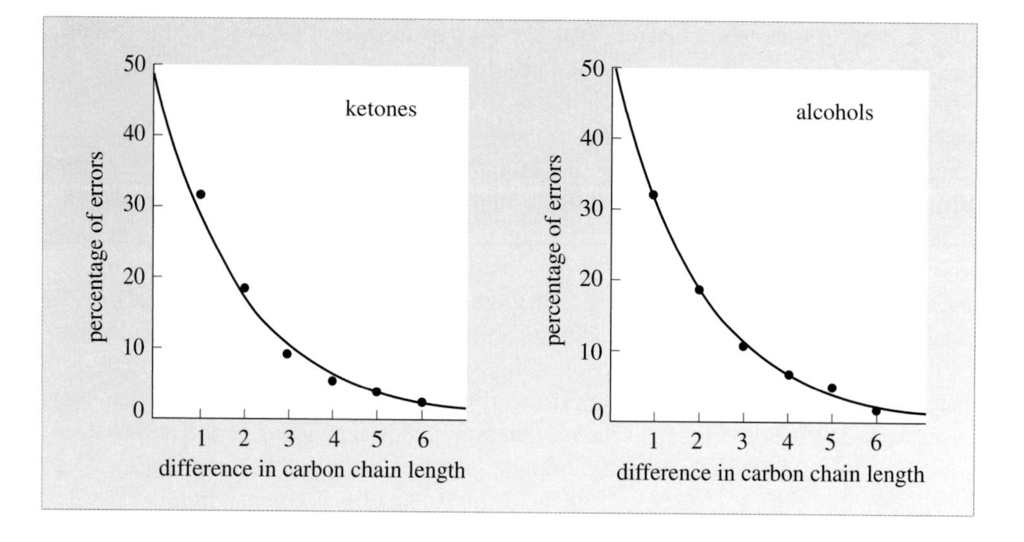

Figure 2.38 Discrimination performance, expressed as the percentage of errors in choosing the 'odd' stimulus, of 20 human subjects as a function of the difference in carbon chain length of ketones and alcohols.

We have already discussed (Section 2.3) how an important molecular feature that is responsible for an odorant binding to an olfactory receptor protein is the functional group. Can functional group differences be distinguished at the level of odour perception? This has also been investigated using the forced-choice triangular test.

❍ Use the structures of, say, ketones, aldehydes, alcohols and acids in Figure 2.36 to design a forced-choice triangular experiment that will allow you to test whether or not functional group structures can be differentiated.

● The experiment should be set up in such a way that only the functional group is varied, which means keeping the size of the carbon chain the same. So a four-carbon ketone ($n = 1$) should be tested in turn against a four carbon aldehyde ($n = 2$), a four-carbon alcohol ($n = 3$) and a four-carbon acid ($n = 2$). The aldehyde, alcohol and acid can also be mutually tested against each other.

The outcome of precisely such an experiment is illustrated by Figure 2.39.

Here, the number of times an odorant pair was correctly discriminated (expressed as a percentage) is shown for each odorant pair investigated. Clearly, all pairs of odorants are able to be discriminated, but some combinations are easier than others. Whereas ketones and acids (pair C–D) can be discriminated with a relatively small error, discrimination between alcohols and aldehydes (A–B) and between aldehydes and acids (B–D) is rather more difficult. Indeed, it seems to be a general observation that ketones and acids are somewhat more distinct in their odour qualities than are the analogous alcohols and aldehydes.

Discrimination between odorants containing different types of functional group is found to be slightly more difficult for molecules that have longer carbon chains as opposed to those that contain shorter chains. You might have expected this to be the case, since we've already discussed the fact that carbon chain length is able to contribute to odour perception. For molecules with longer chains, this contribution begins to mask the contribution of the functional group.

As well as chain length and functional group, a further feature that is important to the properties of molecules is chirality (see Box 2.3 overleaf). Is chirality important to odorant recognition? Interestingly, there is a well known common example of two

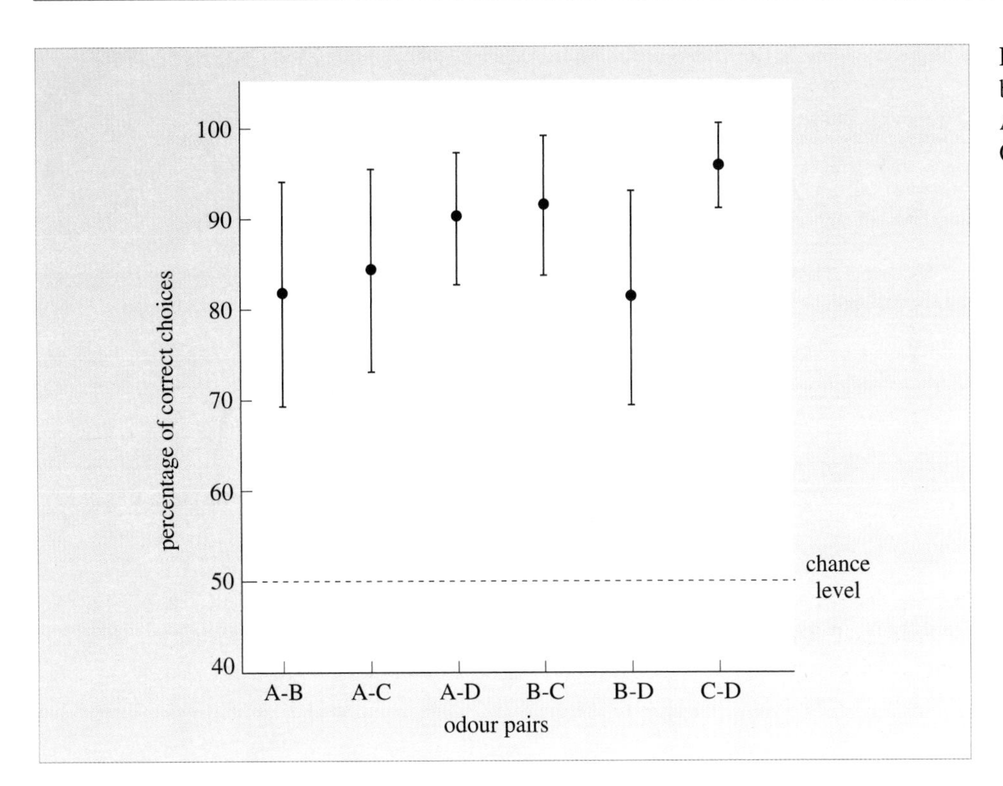

Figure 2.39 Discrimination between odorant pairs: A = alcohol; B = aldehyde; C = ketone; D = acid.

forms of a chiral odorant that are perceived as distinct odours: one form of carvone (Figure 2.40) has a spearmint odour while the mirror image form smells of caraway.

But is this ability to perceive distinct odours for mirror image molecules a general property of the human olfactory system or is it specific to particular molecules? Surprisingly, this question has only recently begun to be addressed. When ten mirror image pairs of odorant molecules were investigated using the forced-choice triangular test (Figure 2.41, overleaf), only in three pairs – α-pinene, carvone and limonene – could the mirror images be discriminated; for the remaining seven pairs – citronellol, menthol, fenchone, rose oxide, camphor, α-terpineol and 2-butanol – selection of the 'odd' odour was only just above the level of chance.

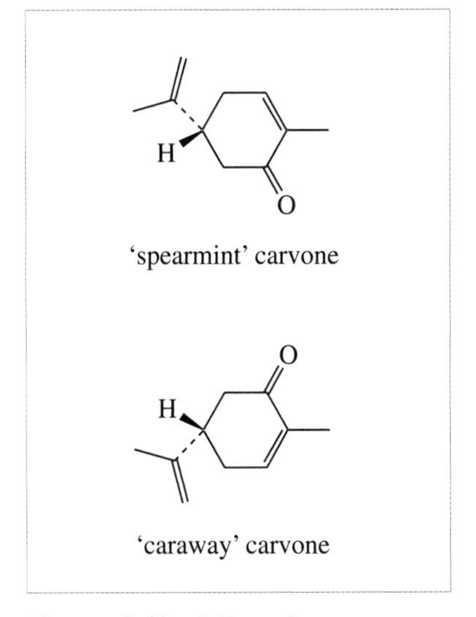

Figure 2.40 Mirror image carvone structures.

❍ Inspect the molecular structures of the chiral odorants in Figure 2.41. Is it possible to discern any molecular features that may give rise to mirror images that can be distinguished and those that cannot?

● There appears to be no discernable pattern between those structures that can be discriminated and those that cannot. Carvone and limonene share several common features (cyclic ring of six carbon atoms, C=C at some position in the ring, CH_3 at same position on the ring, the group opposite the CH_3 at the same position of the ring) but not all of these are present in α-pinene and some are also present in the chiral molecules that can't be discriminated.

This finding is rather surprising. Like all receptors, the olfactory receptors are chiral molecules themselves and their molecular interactions with mirror-image odorant molecules should, in principle, be selective for one of mirror-image forms. It would appear, however, that this is not manifest at the level of odour perception. Why this might be so is not yet clear. One rationale is that the human capacity for chiral

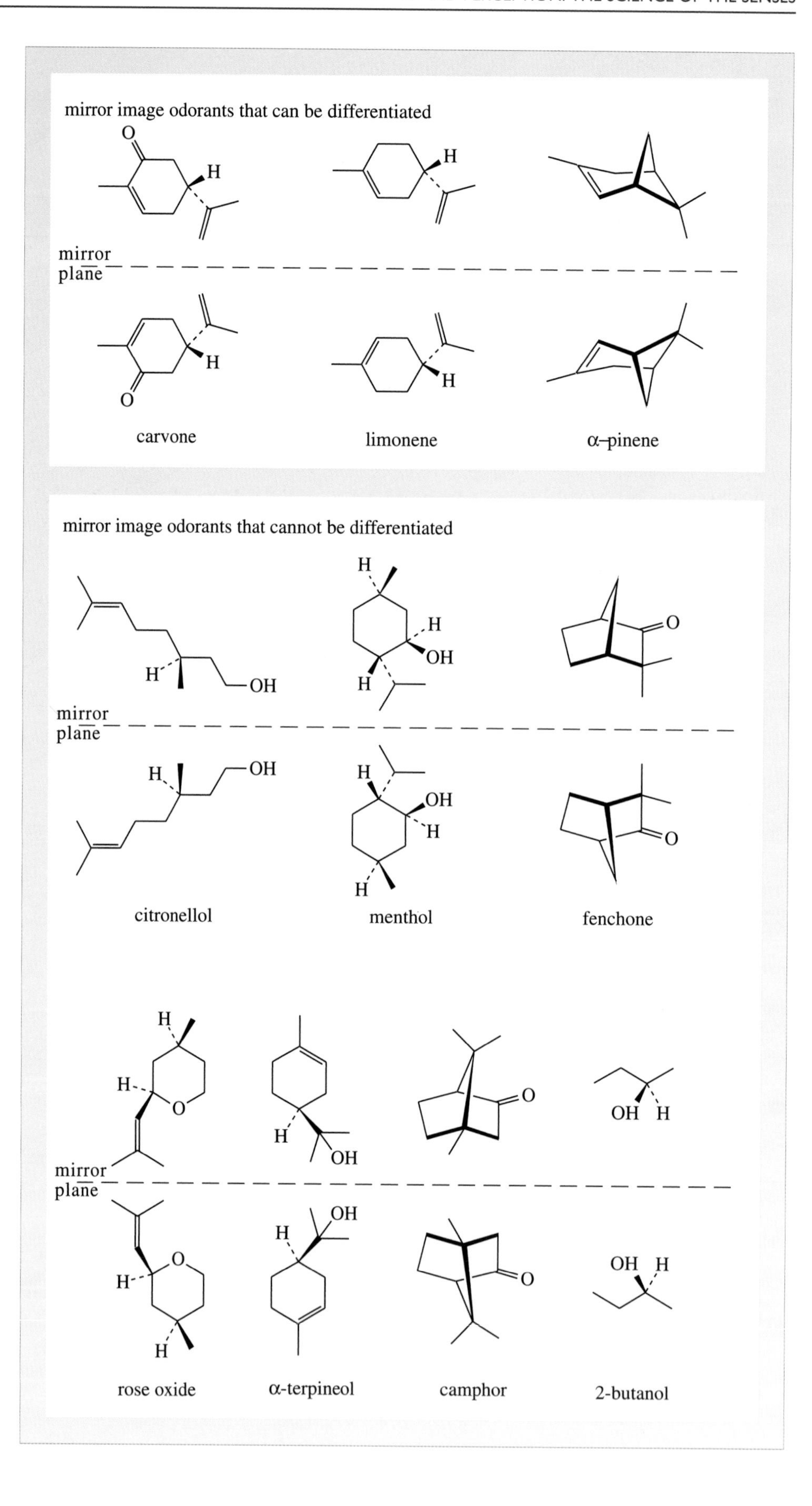

Figure 2.41 Mirror image odorant pairs.

Box 2.3 Chirality

In biological systems, one of the most important aspects of (bio)chemical activity is what is called **chiral recognition**. The term chiral comes from the Greek for hand, *chiros*, and certain types of molecule possess the property of 'handedness', that is, they are mirror images of each other that cannot be superimposed upon each other. Molecules that possess this property contain centres – usually carbon atoms – bonded to *four different* groups that are spatially positioned at the corners of a tetrahedron (all carbon atoms that are bonded to four other atoms always have those atoms positioned tetrahedrally around the carbon). Although this property of handedness is manifest in molecules, it is entirely due to the geometry of the tetrahedron and not peculiar to the molecular world! Figure 2.42 shows two carbon centres that are mirror images of each other. These mirror images cannot be superimposed on one another and are therefore different. For example, if the right-hand image is rotated about the C–W axis a little so that the Y is in the same position as the Y in the left-hand image then the W and Y positions in both images are superimposed. However, the X position in the left-hand image is where the Z is in the right-hand image. In fact, no amount of manipulation will allow all four positions in the two images to be superimposed at the same time.

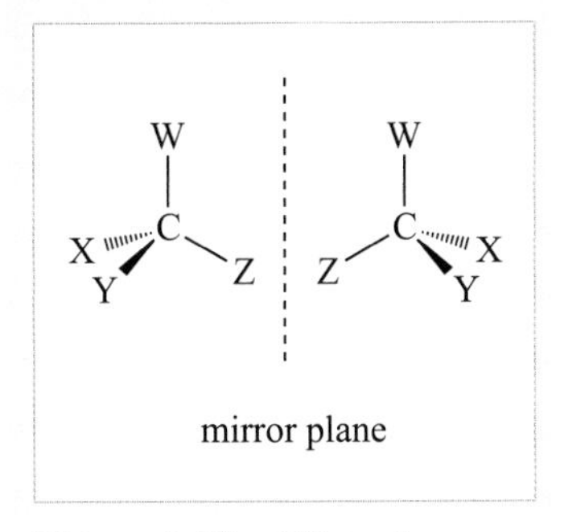

Figure 2.42 Mirror image chiral carbon centres.

Chiral recognition is the ability to distinguish between a molecular structure and its mirror image. In biological systems chiral recognition is ubiquitous, with mirror image molecules generally having different chemical and biological properties. This is particularly true of drugs: for example, one mirror image form of ibuprofen (Figure 2.43) has anti-inflammatory effects while the other form is inactive; and one form of thalidomide has sedative effects but the other is teratogenic, causing malformations in the developing foetus when taken in pregnancy.

active ibuprofen molecule

inactive ibuprofen molecule

active form of thalidomide

teratogenic form of thalidomide

Figure 2.43 Chiral forms of ibuprofen and thalidomide.

recognition is restricted to those mirror image molecule pairs that are found widely in nature. For example, both mirror image forms of limonene, carvone and α-pinene are present in a wide variety of plant extracts, though the ratios of the two forms of each varies according to the source. For menthol, however, it would appear that only one of the mirror image forms is prevalent in all essential oils. This argument would imply that chiral recognition of specific mirror image molecules is a result of evolutionary pressure rather than a general feature of the olfactory system.

Our discussion so far has focused on those factors that affect our ability to discriminate one odorant from another. However, seldom are we presented with an odour that is due to one unique odorant. Whether it is a flower fragrance, a perfume or the unpleasant smell of rotting vegetables, the smell we perceive is a combination of many different odorant molecules. How good is our discriminatory ability at trying to identify a particular odorant when it is present in a mixture? And is the combination of odours perceived as a 'new' odour entirely? The consensus seems to be that, in answer to the first question, we find it difficult to identify more than three odorants, even when the mixture only contains four, and in answer to the second, we appear to synthesise no new odour 'qualities' that are not already expressed by the mixture components themselves.

For example, participants in an odour identification study were familiarized with the four odours involved (Figure 2.44). Then they were presented randomly with the four single odorants, the six possible two-odorant mixtures, the four three-odorant mixtures and the mixture containing all four odorants and asked to identify whether or not one particular odorant was present. The experiment was repeated three more times, each time asking the participants to identify the presence or absence of a different odour molecule. The results (Figure 2.45) are illuminating!

Correctly identifying the source when exposed to single odorants is straightforward (Figure 2.45a). Even with mixtures of two odorants it proved possible to easily identify, separately, the components present. With three-odorant mixtures however the ability to identify any one of the odorants is reduced significantly, and with the four-component mixture no individual odorant could be identified above the level of chance. It would appear that, for these particular odour molecules at least, humans begin to experience difficulty in identifying individual odorants in mixtures that contain as few as three components.

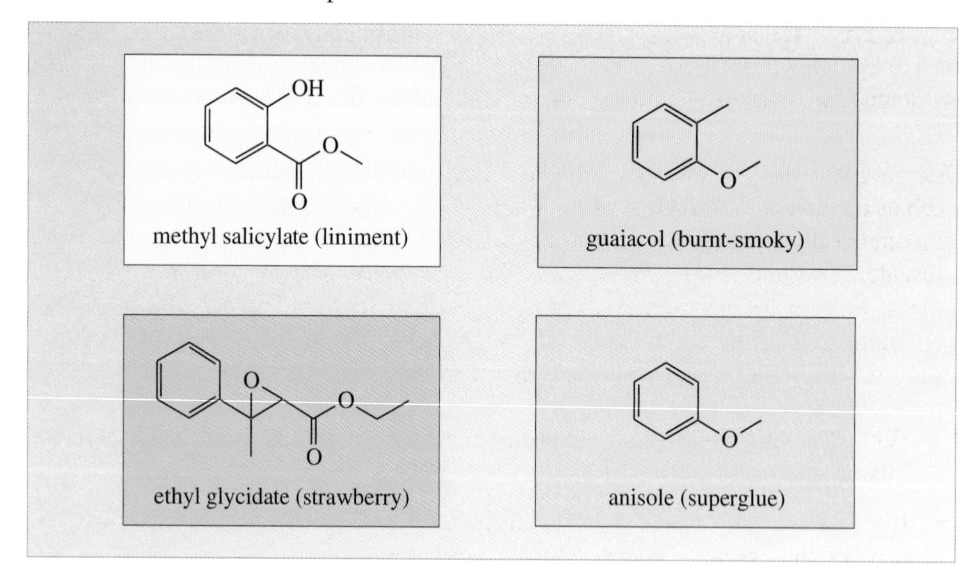

Figure 2.44 The molecular structures of four odorants used in an odour identification experiment.

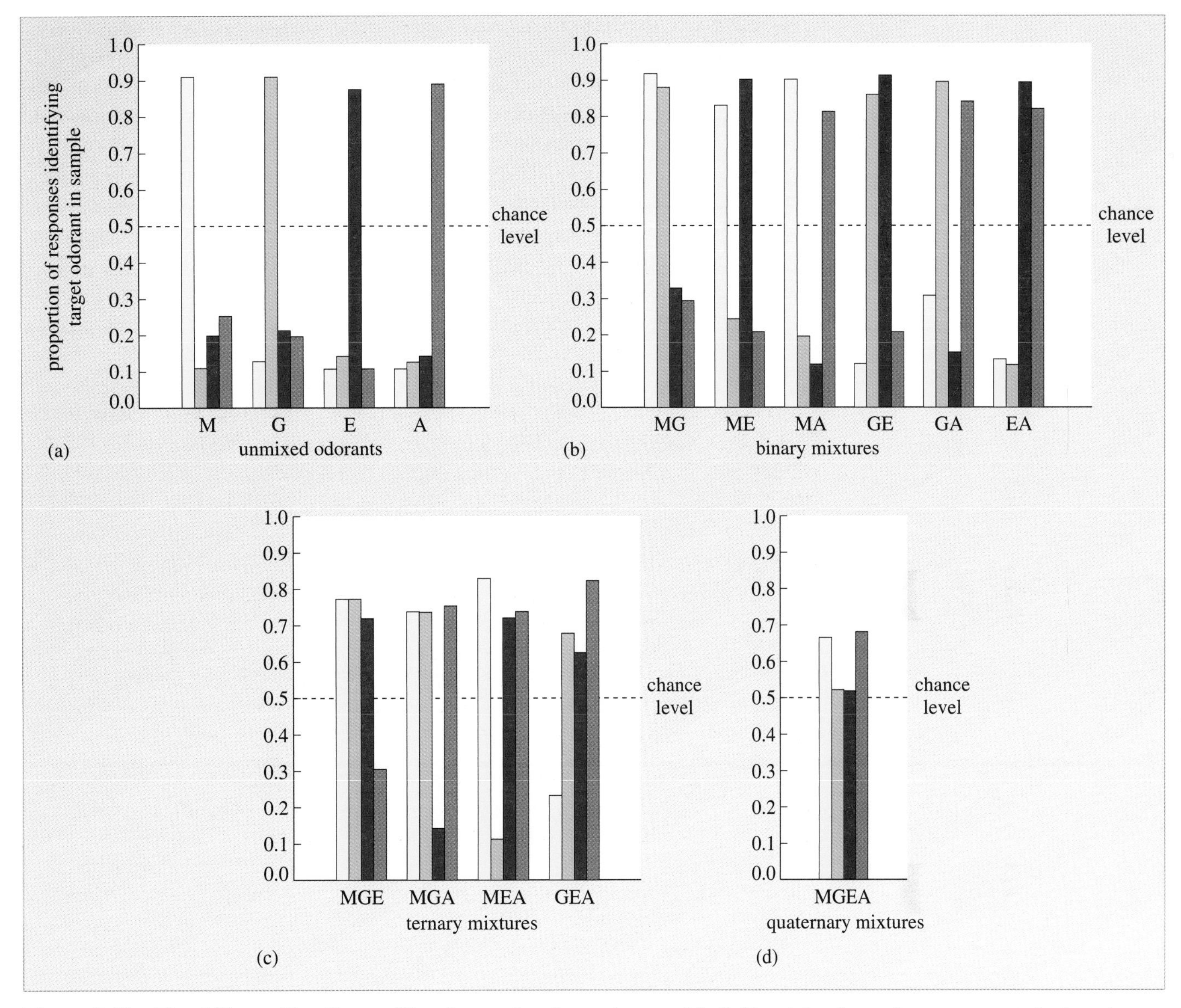

Figure 2.45 The ability to identify specific odorants in odour mixtures. M, G, E and A refer to the components in the odours: M, methyl salicylate; G, guaiacol; E, ethyl glycidate; A, anisole. The colours of the histogram bars are the same as those used to identify the structures in Figure 2.44. So, yellow represents methyl salicylate, etc.

The participants were asked to describe the odours they perceived using words – such as medicinal, eucalyptus, fruity, minty – from a 146-word list. One surprising outcome of this second experiment was that words that were unique to certain individual odorants, for example 'burnt' for guaiacol or 'strawberry' for ethyl glycidate, were not used at all to describe even the two-component mixtures containing these compounds. Yet Figure 2.45b shows that these compounds were easily recognized in such two-component mixtures.

❍ What does this imply about our ability to perceive and identify an odorant in a mixture?

● It seems to imply that we can use other, less prominent components of the odour perception to recognize an odorant in a mixture.

A further observation of this 'descriptive' study was that the most common words chosen to describe the mixture containing all four odorants included nine of the twelve major descriptors of methyl salicylate. Even so, methyl salicylate was not recognized as a component of the mixture above chance level. Why this is so is currently not clear. One interpretation is that odorant identification requires an ability to discriminate the *relative* perceived intensity of its 'descriptors'. In a pure compound, the relative intensities of these qualities are fixed. But when these relative intensities are altered, as they can be when the odorant is present in a mixture, the overall odour perception may be sufficiently different from the individual odorants for them not to be recognized. Whatever the reason, at no point in this experiment was the perceived odour described using words that were different from those used for the individual odorants.

Such studies have led to a **configurational hypothesis of olfaction** being proposed. In the same way that the individual components of a face – eyes, nose, mouth, etc. – can be discerned and then perceptually fused into a recognizable face, so the configurational hypothesis of olfaction implies that the odorant profile is discerned first, followed by a processing of the relationship between the components of the profile before the overall odour perception is experienced. In simple mixtures, individual odorants can be identified because their individual profiles remain largely intact. In more complex mixtures, the features of each odorant profile are sufficiently altered such that individual components are not identifiable. In these circumstances we tend to associate the resultant profile with the odour source, for example chocolate or a brewery. There is no one molecule that smells like either of these aromas. Rather, these aromas arise from a synthesis of the features of the many odorant molecules that make up these complex multicomponent mixtures.

Of course, it could be argued that these conclusions have arisen from studying only four molecules that have rather similar molecular structures and that this similarity lies at the root of the poor olfactory ability of humans to discriminate between them. However, largely similar conclusions have been reached from using larger odorant sets, each containing eight odorants that were selected by a panel of perfumers and flavourists. In one set, the odorants were selected on the basis that they are easily identified in their own right but blend well when mixed with other odorants so are difficult to discriminate in mixtures. The second set contained odorants that were selected because they have distinctive odours that remain identifiable in mixtures, that is they don't blend well. These two sets are shown in Table 2.4 and it is clear that their odours and molecular structures generally have little in common.

Even so, the human ability to identify the components of a mixture of the 'good blender' odorants or of the 'poor blender' odorants was found to be remarkably similar. This is shown in Figure 2.46 (overleaf), which illustrates the ability to correctly identify all of the odours in a mixture as the complexity of the mixture increases.

For mixtures that contain fewer than four odorants, it would appear that the human ability to discriminate the individual 'poor' blenders is slightly better than for the 'good' blenders, though the differences between the two sets are not great. As you might expect, for either set the ability to identify the components is higher the less complex the mixture.

Table 2.4 Odorants used to examine the human capacity for odour discrimination.

Odorant	Odour description	Molecular structure
Good blenders		
phenethyl alcohol	rose	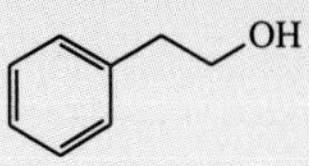
exaltolide	musk	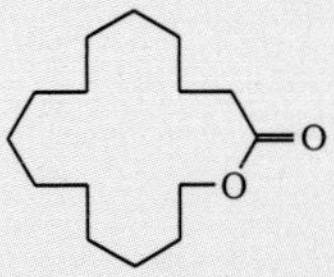
cinnamaldehyde	cinnamon	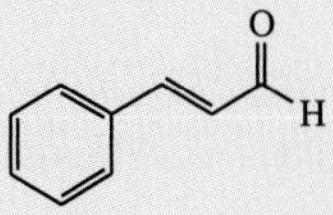
γ-nonalactone	coconut	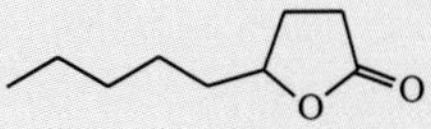
ethyl butanoate	fruity	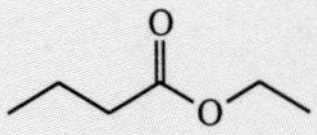
(+)-limonene	orange	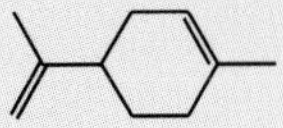
furaniol	burnt caramel	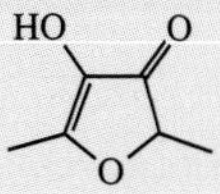
benzaldehyde	almond	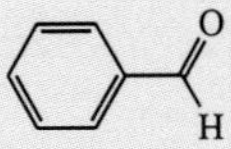
Poor blenders		
skatole	bad breath	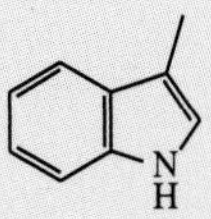
1-octen-3-ol	mushroom	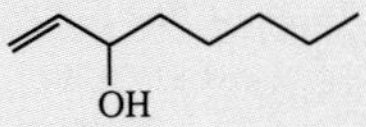
cis-3-hexenol	cut grass	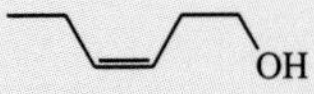
methyl salicylate	liniment	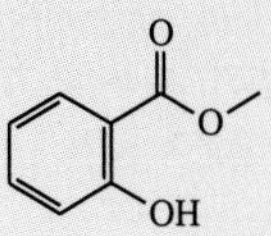

Table 2.4 *continued*

Odorant	Odour description	Molecular structure
Poor blenders		
diallyl sulfide	garlic	
2,5-dimethylphenol	antiseptic	
styrene	glue	
1-dodecanal	mandarin	

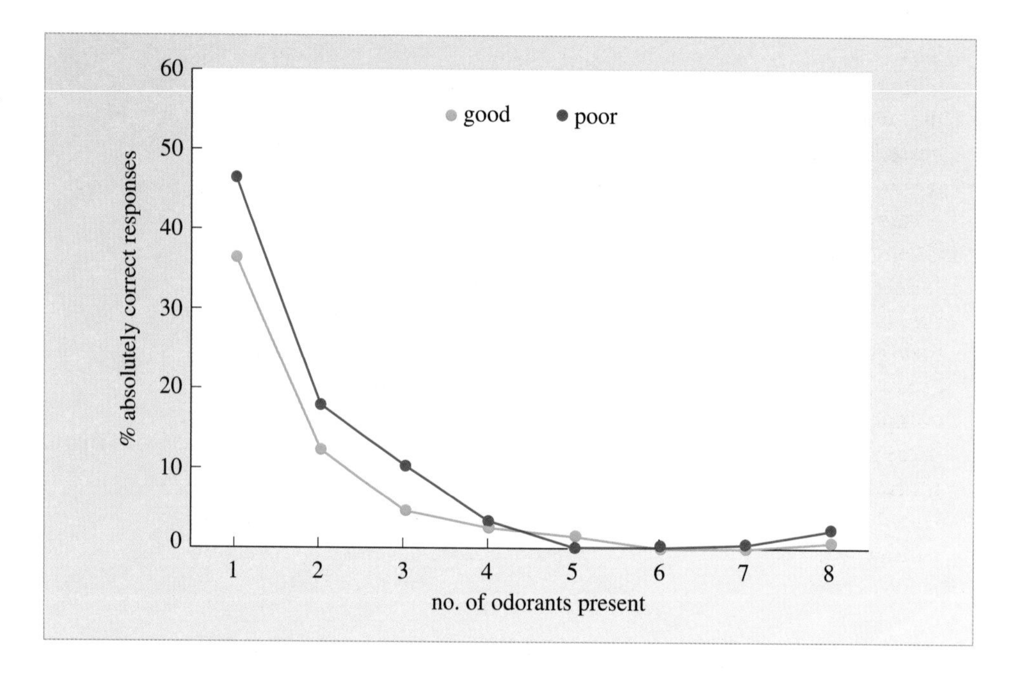

Figure 2.46 The ability to identify all the components in odorant mixtures involving 'good' and 'poor' blender molecules.

❍ From Figure 2.46 what two deductions can you make about the human ability to identify the individual components in mixtures containing four or more odorants?

● First, the ability to identify all the components in mixtures containing four or more odorants is very low (less than five per cent). Second, there appears to be no difference between the ability to recognize the components of mixtures that involve odorants that blend well and odorants that blend poorly.

Once again, the study implies that the human ability for odour recognition of individual components of mixtures is largely restricted to three, or at best four, whatever the molecular structure or odour quality of the individual components. This limited range of odorant recognition is surprising given that we exist in an environment that contains an enormous number of complex odorant mixtures which must be discriminated from each other. This would tend to imply that the human olfactory system:

> … has evolved to discriminate a small number of relevant stimuli, and to simplify the mass of olfactory information that is available … rather than be broadly responsive to a large range of odours.
>
> A. Livermore and D. G. Laing (1998) *Physiol. Behavior*, **65**, pp. 311–320.

This is probably why benzaldehyde and hydrogen cyanide both have an almond smell (Box 2.4). In this way, the limited human capacity for recognizing individual components of an odour mixture is the 'downside' of a sensory system that provides a rapid, accurate and efficient means of identifying environmentally relevant (food, danger, etc.) complex mixtures.

Box 2.4 The smell of bitter almonds

Why do benzaldehyde and hydrogen cyanide both smell of almonds?

One famous problem in the correlation between odour and molecular structure is that of the smell of bitter almonds. Benzaldehyde and hydrogen cyanide are the two compounds that best elicit this odour. The more a molecule resembles benzaldehyde in terms of its size, shape and charge distribution, the more likely it is to smell of almond. But, as noted earlier, the structure of hydrogen cyanide is quite unlike that of benzaldehyde. So why do they have the same smell? Clearly, their almond odour cannot be explained by assuming that there is an 'almond' odorant receptor. However, if the almond odour 'exists' only in the brain as a result of the pattern of nerve signals elicited by the molecules, then it becomes possible to appreciate why they might have the same odour. In almonds, benzaldehyde and cyanide always occur together. They are components of a compound, synthesized by the plant, called amygdalin that breaks down to form a molecule each of benzaldehyde, hydrogen cyanide and glucose. Consequently, it is likely that the brain will have learnt to interpret the signals arising from benzaldehyde and those from hydrogen cyanide in a similar way.

You should now read Chapter 24 of the Reader, *The perception of smell* by Tyler Lorig.

Activity 2.3 Smell

Now would be a good time to work through the smell section of the 'Smell and Taste' sequence on *The Senses* CD-ROM. This provides a useful overview of the olfactory system. Further instructions are given in the Block 6 *Study File*.

2.13 Summary of Sections 2.10–2.12

The response of olfactory receptors is seen to saturate with odorant concentration, giving rise to a maximal effect that can be obtained. The odorant concentration at which a receptor saturates can vary from receptor to receptor, so it is possible that some receptors are used as concentration detectors for odorants. Additionally, as the odorant concentration increases, a different pattern of receptors can be activated – an effect also seen at the glomerular level – which could account for the changing odour perception of some odorants with concentration.

Smell is characterized by an adaptive response, the odour perception diminishing upon prolonged exposure to it. There are several types of adaptation seen in the olfactory system. Short-term adaptation occurs upon exposure to very short bursts of odorant exposure. Odour response desensitization occurs to longer exposures to odorant, and long-term adaptation is the adaptive response that lasts on the minute timescale after even very short odorant exposures.

The human ability to discriminate between odours seems limited. It appears that we find it difficult to distinguish between similar molecules of different sizes unless they differ by three or more CH_2 groups. We are better at discriminating between molecules of similar size but with different functional groups, though certain functional groups (e.g. ketones and acids) are easier to discriminate than others (e.g. alcohols and aldehydes). The human olfactory system can distinguish between chiral molecules, but this does not seem to be a general property of the system. Most likely, for those chiral molecules that can be distinguished there has been an evolutionary pressure for humans to recognize the difference between the two molecules.

When it comes to recognizing individual odorants in mixtures, we have difficulty when the mixture contains three or more components and are unable to do so above the level of chance when the mixture contains four or more components. This perception difficulty appears to be independent of molecular complexity or the odour of the individual odorants.

Question 2.1

Which of the following molecules would you predict might be odorants:

$C_2H_4O_2$ (b.t. 118 °C); C_2H_5NO (b.t. 221 °C); $C_3H_8O_3$ (b.t. 290 °C); $C_5H_8O_2$ (b.t. 139 °C)?

Question 2.2

Which of the receptors in Figure 2.13 are most tuned to (i) molecules containing nine carbon atoms, (ii) carboxylic acids, (iii) alcohols and (iv) longer carbon chains.

Question 2.3

(i) Use the *Olfactory receptors* exercise on the CD-ROM to complete the following table that shows which of the receptors A–D, and with what strength of response, are activated by the odorants 1–6. The first row is completed as an example.

(ii) What pattern of receptor activity would be obtained by a mixture of odorants 2 and 3?

Table 2.5 The activation of receptors by odorants.

Odorant	Receptor			
	A	B	C	D
1	s†	w		vw
2				
3				
4				
5				
6				

† vw = very weak, w = weak, m = medium, s = strong, vs = very strong.

Question 2.4

From what you know about the mechanism of olfaction and coding, why might it be easier to differentiate between odours of butanol and octanol than it is between the odours of heptanol and octanol?

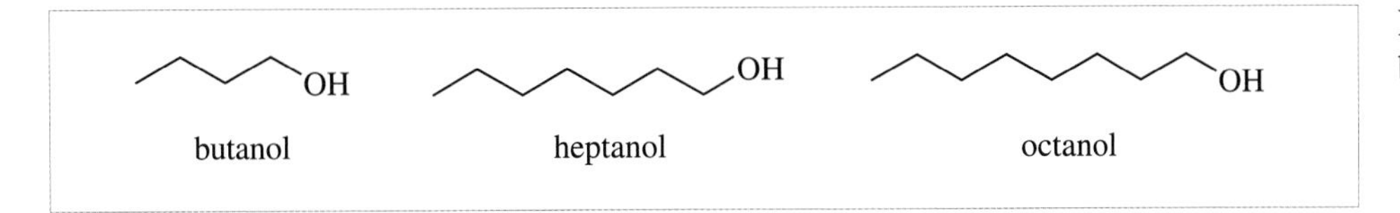

Figure 2.47 The structures of butanol, heptanol and octanol.

Question 2.5

Why does someone suffering from body odour not seem to notice while those nearby do?

Question 2.6

The concept of the olfactophore – a pattern of chemical bonding that gives rise to a particular odour quality – was useful when it was thought that particular receptors were responsible for particular odours (Amoore's theory, a version of the labelled-line hypothesis). Why do you think the olfactophore concept remains useful, for example, to the perfumery industry now it is known that odour coding is a combinatorial process?

Question 2.7

Assume (i) there are five olfactory receptors *a*, *b*, *c*, *d* and *e*, (ii) that to elicit an odour at least three receptors must be activated, and (iii) that to discriminate between odours the odour codes must be different by at least two activated receptors. How many different odours can be detected, and how many discriminated?

Question 2.8

From your study of this course, what do you think are the main factors that contribute to odour perception?

Question 2.9

Why do different odorants have different detection thresholds?

Question 2.10

When the olfactory system is repeatedly exposed to the odour of lemons, the subsequent perception of limes is altered significantly but there is no change to the perception of peanut butter. Suggest why this may be so.

3 Making sense of taste

In *La Physiologie du Goût*, published in Paris in 1825, the French gastronomist Jean-Anthelme Brillat-Savarin wrote that:

> … of all our senses, taste, such as Nature has created it, remains the one which, on the whole, gives us the maximum delight: […] because, when we eat, we experience an indefinable and peculiar sensation of well-being, arising out of an instinctive awareness that through what we were eating we are repairing our losses and prolonging our existence.

It is not difficult to appreciate this connection between our sense of taste and the need to 'repair our losses and prolong our existence'. If you are weary, perhaps because you haven't eaten for some time or because you have been on a long walk, you may experience a craving for sweet foods, or if you have undertaken more 'violent' physical exercise you may have found yourself needing to eat salty and savoury foods. A possible interpretation of these differing responses is that in the former you need to replenish energy stores whereas in the latter it is the salt, lost through sweat, and amino acids that need replacing. However, one needs to exercise care in rushing to these conclusions; while carbohydrates are a valuable source of energy, not all carbohydrates are sweet like sugar, and not all sweet foods are good energy sources. Nevertheless, there does seem to be a close connection between our sense of taste and our nutritional requirements.

However, before we proceed any further with our discussion of taste we should set out what we mean by this particular sense. In everyday language we often use the word 'taste' to mean the sensation that we get when we place materials, usually food, into our mouths. Used in this way, taste has associated with it our sense of smell. In fact, it has been estimated that olfaction contributes up to 90 per cent of our sensory appreciation of food. It is easy to demonstrate this by, for example, eating a banana while pinching one's nostrils. This prevents any volatile odorants in the banana from entering the nasal cavity from the back of the mouth and stimulating our olfactory system. Eaten this way, the banana 'tastes' rather bland. We get a similar experience if we have a heavy cold and our nasal passages are blocked with mucus. The 'overall' sensation, including the sense of smell, that we perceive when eating food, is better termed as *flavour* (and we shall return to that in Section 4), while the sensation we perceive when olfaction is absent is more appropriately termed **gustation** (though the terms gustation and taste, when it is meant in this restricted sense, are often used interchangeably).

❍ Try to identify some common substances in your home that have taste but no smell. What tastes do you associate with them? How many different types of taste can you recognize?

● Two very obvious examples come to mind: sugar, which tastes sweet, and salt which tastes salty. Other sweet tasting materials that have no smell are saccharin and glycerol (glycerine).

Actually, it is very difficult in our everyday lives to encounter many substances that have no smell, largely because most materials we encounter are complex mixtures, some of which are volatile and stimulate our olfactory system. As it happens, all three of the above examples – sugar, which is the chemical sucrose, saccharin and salt, which is the chemical sodium chloride – are single compounds.

So there are at least two different types of taste sensation. But are there others, and if so how many? If you have some carbonated water to hand you may like to taste that alongside a non-carbonated equivalent. Can you detect a taste to the carbonated water? You will find that it has a slightly sour taste. You get a similar sensation if you taste a weak vinegar solution while pinching your nose, or from vitamin C (provided it hasn't had any flavourings added). Plain yoghurt is a complex mixture of materials; it, too, has a decidedly sour taste. If you've ever taken aspirin for a headache then you will have experienced a fourth type of taste sensation. Aspirin is decidedly bitter, so much so that you may only have been able to tolerate it as a child if it were crushed up in some jam. Incidentally, aspirin was synthesised as a more palatable alternative to the naturally occurring, and very bitter, anti-inflammatory compound salicylic acid! Alternatively if you enjoy tonic water, then you will have experienced another bitter tasting substance – quinine, a drug that has use in the treatment of malaria.

That makes four different types of taste experience: sweet, salt, sour and bitter. For many years these were considered the four 'basic' tastes. In recent years a fifth taste, **umami**, has been added to the repertoire. Umami is a Japanese word that has no satisfactory English translation, but is often described as 'savoury'. It is associated with the taste experience brought about by monosodium glutamate (MSG), a component of partially fermented foods like soy sauce.

Although in our exploration of the sense of taste we will invoke this commonly used division into five 'basic' categories, it is worth recognizing that such a description is not universally accepted as helpful or even necessary. For example, subjects were asked to taste a range of concentrations of a variety of pure substances – ranging from common taste stimuli such as sodium chloride ('salt'), citric acid, caffeine, sucrose ('sugar') and monosodium glutamate to non-traditional taste stimuli such as lithium chloride, potassium chloride, aspartame (Nutrasweet®), sodium benzoate, 5′-inosine monophosphate and 5′-guanosine monophosphate – and then to sort them into groups of the same taste quality. At first, each individual was allowed to sort without being provided with any labels for the groups, but they were allowed to generate their own labels for the groups they had formed. This was to prevent the subjects having any preconceived ideas about what groups were 'required'. In this 'open' situation, the number of different labels matched the number of groups; that is, each group was given a separate label. Then the participants were asked to re-label the groups using combinations of the terms sweet, bitter, salty, sour and 'other'. In this 'closed' situation, for approximately forty per cent of the participants, more than one group was given the *same* label such that there were fewer labels than groups. So, when prompted to sort using preconceived categories, the number of possible taste qualities can be significantly underestimated. Indeed, by restricting the choice of descriptors or the tastes employed in the test, taste description according to the five 'basic' categories can become circular and self-fulfilling. As American psychologist Jeannine Delwiche has written:

> … it is wiser to accept the multiplicity of taste sensation without relying on a flawed organizational structure.

It seems sensible to acknowledge this caveat, while also recognizing that our current understanding of taste is organized around the five broad descriptive categories.

3.1 The taste machinery

The anatomy of the gustatory system is dealt with in more detail in Chapter 25 of the Reader and on *The Senses* CD-ROM. However, we shall briefly review the system here to aid your study of the remainder of this block.

It is fairly obvious that the detectors of the gustatory system are housed in the oral cavity. The detection units are the taste cells. These are specialized epithelial cells that are housed in **taste buds** located on the tongue and also on the soft palate. You might want to try carefully rubbing a small cotton wool swab that has been soaked in a salt solution along the roof of your mouth to verify that you have taste buds on your palate. On the tongue, taste buds are part of larger structures called **papillae**, and each papilla contains several taste buds. These papillae are the tiny projections that give the tongue its uneven appearance. There are four types of papillae (Figure 3.1 overleaf) – filiform, fungiform, foliate and circumvallate – but only the latter three house taste cells. Fungiform papillae appear as pinkish spots on the front part of the tongue and are easily visualized if you place a drop of blue food colouring on your tongue. The circumvallate papillae are roughly a dozen in number, and are the ones that form an inverted V at the back of the tongue, while the foliate papillae form small trenches on the sides of the rear of the tongue. Humans have about 9–10 000 taste buds in all, of which approximately 20 per cent are in the fungiform papillae, 30 per cent in the foliate papillae, 25 per cent in the circumvallate papillae, 15 per cent in the palate and 10 per cent on the laryngeal surface of the epiglottis. Taste buds are segmented structures, each of which can contain between 50 and 100 taste cells. The taste cells themselves have thin projections – called microvilli – that project from the top of the cell into the taste pore where they can contact the taste molecules that are dissolved in saliva.

Unlike olfactory receptor cells, which are neurons that send axons to synapses in the olfactory bulb, taste cells do not possess axons. Instead, they synapse with neurons from cranial nerves VII, IX and X. The chorda tympani branch of cranial nerve VII innervates the front part of the tongue, while the back of the tongue is innervated by the glossopharyngeal nerve (cranial nerve IX).

❍ Which taste cells does the chorda tympani innervate, and which the glossopharyngeal nerve?

● The chorda tympani innervates taste cells in the fungiform and the anterior foliate papillae, the glossopharyngeal nerve innervates those in the circumvallate and posterior foliate papillae.

Interaction of a tastant molecule with the taste cell results in a release of neurotransmitter across the synapse to trigger the neuronal impulse. These signals are conducted first to the nucleus of the solitary tract, from there to the thalamus, thence to the primary taste cortex and from there to amygdala and orbitofrontal cortex where they combine with signals from the olfactory, visual and somatosensory systems (Figure 3.1).

As with olfaction, our study of gustation will begin with the molecules of taste, then will investigate the different types of taste modalities and how they are activated, and will be followed by a discussion of our understanding of how gustatory information is coded.

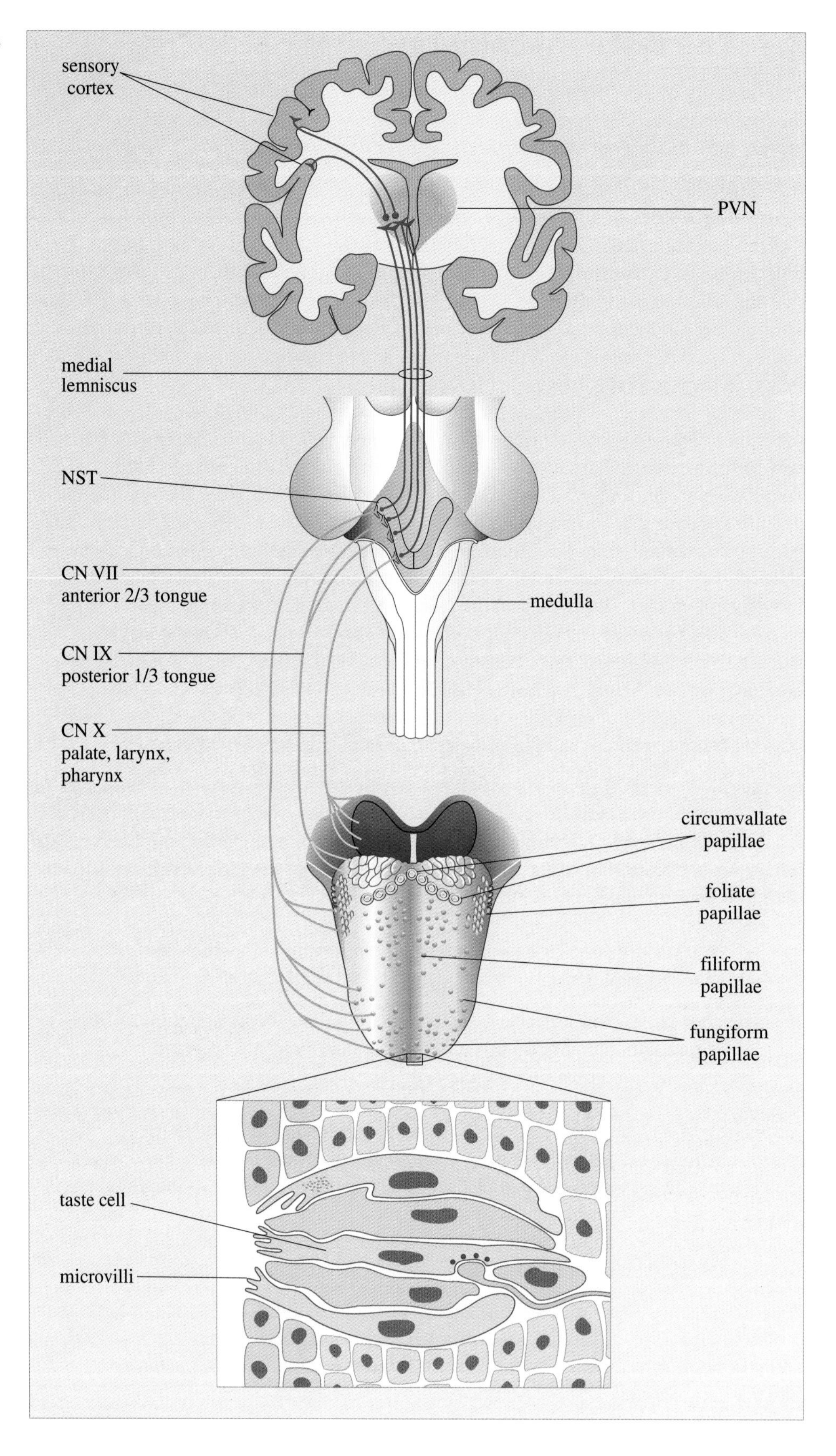

Figure 3.1 The main components of the human gustatory system: CN, cranial nerve; NST, nucleus of the solitary tract; PVN, posteromedial ventral nucleus of the thalamus.

You should now read Chapter 25 of the Reader, *The gustatory sensory system* by Tim Jacob.

Activity 3.1 Taste

At this point you may like to work through the taste section of the 'Smell and Taste' sequence on *The Senses* CD-ROM. Further instructions are given in the Block 6 *Study File*.

3.2 The molecules of taste

In the same way that molecules that bring about an odour perception are called odorants, molecules that bring about a taste perception are called **tastants**. In order to reach the nasal epithelium, odorants must be volatile, a condition that restricts their molecular size. Tastants have a different restriction, but one that does not limit so severely their molecular size. Since tastants are put physically into the oral cavity, to come into contact with the taste cells of the tongue and oral cavity they must be soluble in water to some extent. Now, water is a polar molecule (Figure 2.4) in which the oxygen atom is slightly negatively charged and the hydrogen atoms slightly positively charged, and in which the oxygen atom of one water molecule is hydrogen bonded to a hydrogen atom of another water molecule.

❍ Given that water is polar, what kinds of substances will dissolve in it, and what kinds of interactions will be involved between water molecules and the materials dissolved in it?

● Polar materials tend to dissolve polar substances. The interactions between water and the substances dissolved in it will involve hydrogen bonding and also attraction between the polar parts of the water molecules and oppositely charged polar regions of the dissolved substances.

So, rather than having a size restriction, tastants must have structures that enable them to interact with water molecules using hydrogen bonding, interactions between oppositely charged regions, or a combination of the two. For this reason, tastants have a much more diverse range of structures and sizes than do odorants. For example, sweet tasting molecules vary from those that are small, have low relative molecular masses and are neutral, like glycerol which has a relative molecular mass of 92, through ionic compounds like sodium saccharin, a synthetic sweetener of relative molecular mass of 205, to large, naturally occurring proteins like thaumatin (relative molecular mass 22 209) which has both charged and neutral polar groups at its surface (Figure 3.2 overleaf).

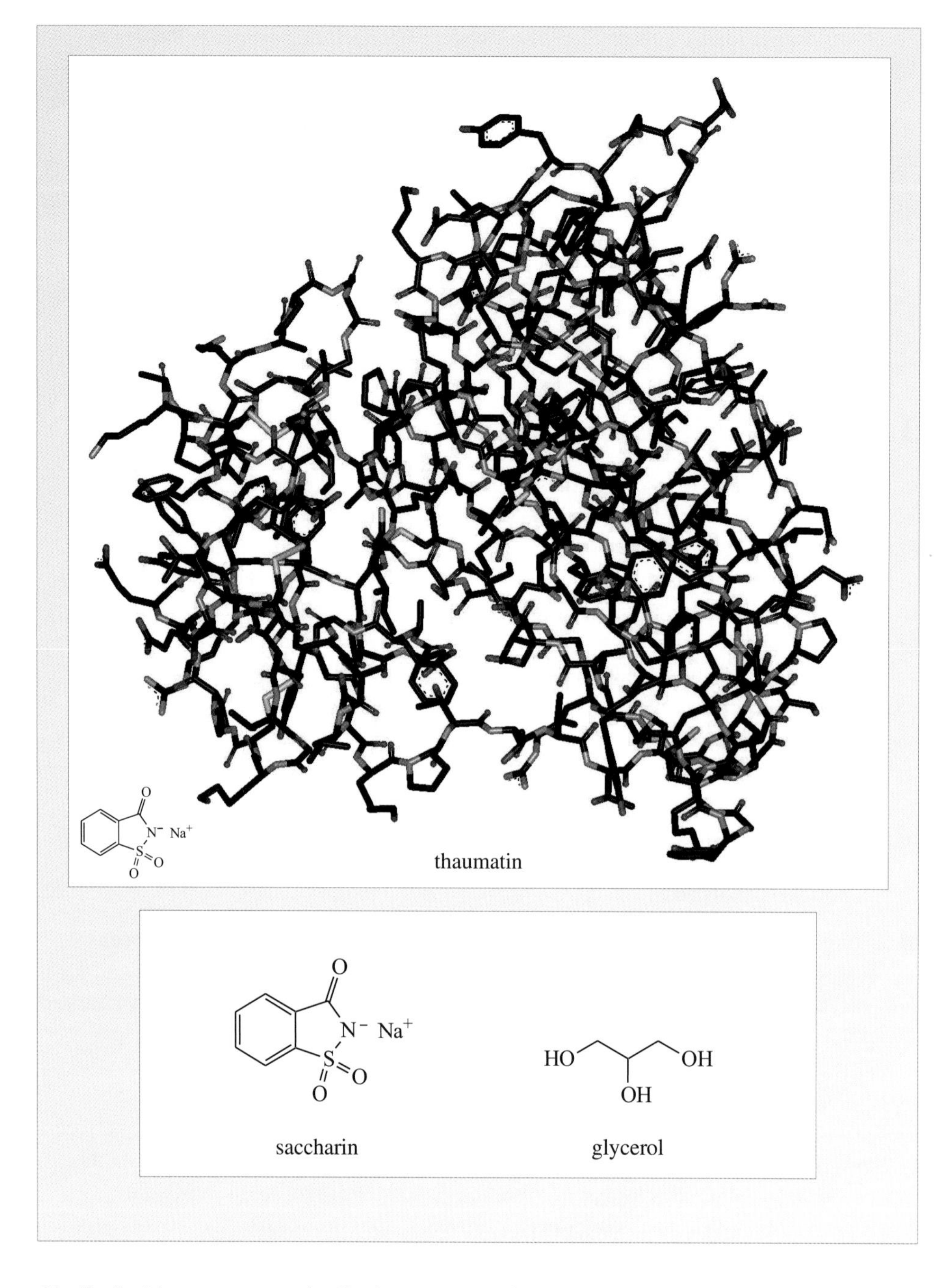

Figure 3.2 Three sweet-tasting molecules: thaumatin, sodium saccharin and glycerol. Sodium saccharin is also drawn alongside, and at approximately the same scale as, thaumatin so that their relative sizes become apparent.

Similarly, bitter compounds display a range of molecular architectures. There is little molecular similarity between aspirin, caffeine and quinine (Figure 3.3), for example, or between any of these and denatonium benzoate, claimed to be the most bitter tasting substance known to humankind. As Bitrex®, the latter is added to many dangerous household products to prevent them from being ingested.

Curiously, some sweet compounds such as saccharin are also said to have a bitter aftertaste, yet saccharin does not closely resemble the structure of bitter compounds. What is more, compounds that have almost identical molecular structures have very different taste perceptions. So, dulcin (Figure 3.4) tastes sweet but its analogue phenylthiocarbamide (PTC), in which the oxygen atom of dulcin is replaced by a sulfur atom, tastes bitter. Similarly, the sugar α-D-mannose is sweet

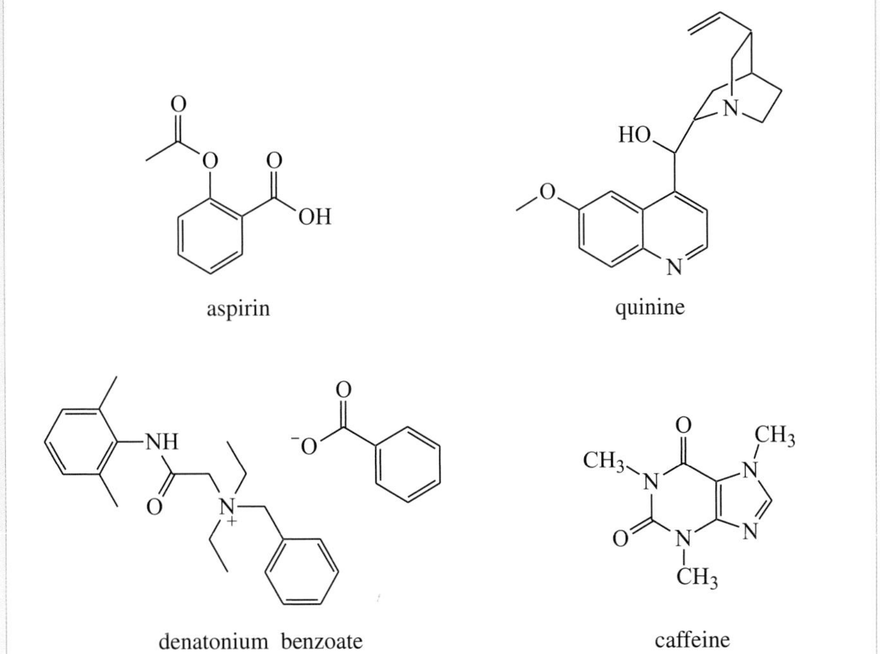

aspirin

quinine

denatonium benzoate

caffeine

Figure 3.3 Four bitter-tasting molecules: aspirin, caffeine, quinine and denatonium benzoate.

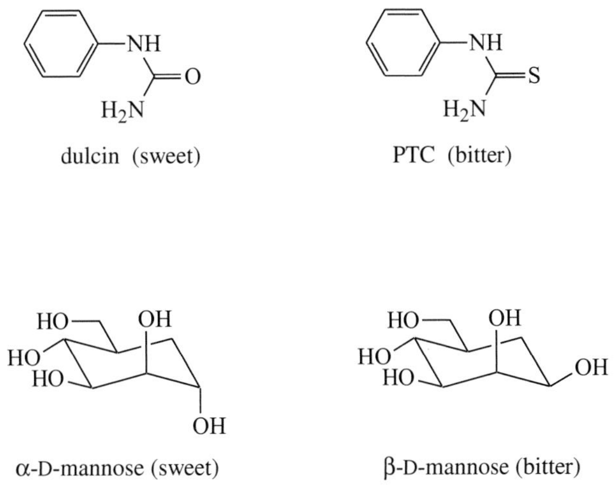

dulcin (sweet)

PTC (bitter)

α-D-mannose (sweet)

β-D-mannose (bitter)

Figure 3.4 Structurally similar molecules that have different taste perceptions.

but β-D-mannose, which has an identical structure except for the position of one of the OH groups, tastes bitter. Identifying the cause of such profound perceptual shifts by minor changes in molecular structure is something that occupies the research efforts of taste chemists worldwide.

Like the molecules that elicit bitter and sweet tastes, the known umami-generating tastants, though currently few in number, also exhibit large structural differences. Some, like glutamic acid, are amino acids; others, like GMP, are ribonucleotides (the building blocks of nucleic acids) (Figure 3.5 overleaf).

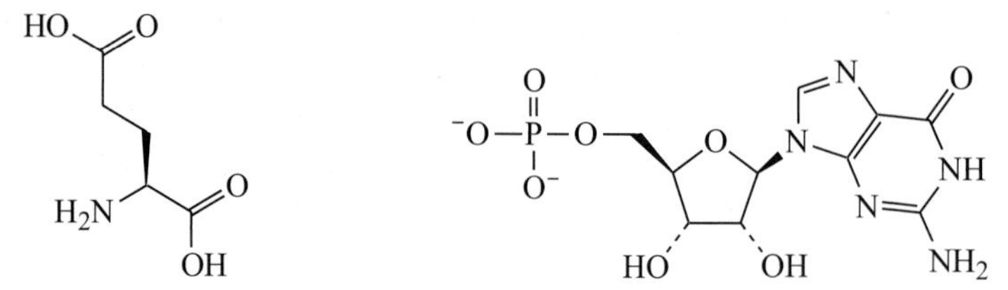

Figure 3.5 Two umami-tasting molecules.

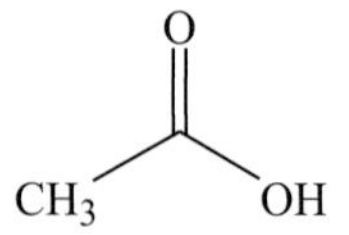

Figure 3.6 Two sour-tasting substances.

Substances that elicit a sour taste are also diverse in structure (Figure 3.6). On the one hand, organic acids such as acetic acid have a pronounced sour taste, on the other hand, so do mineral acids such as hydrochloric acid. Moreover, whereas mineral acids exist as charged ions (H^+ and Cl^-) in aqueous solutions, the organic acids exist largely as the neutral molecule. So how is it that both types of substance elicit the same perception? We shall explore this later.

Finally, except for the sour response to mineral acids, all of the taste mechanisms mentioned so far respond to organic molecules of varying sizes. The salt taste mechanism redresses this balance, because it provides us with the ability to detect inorganic ionic substances. The most obvious example is what we call common salt, that is sodium chloride. This substance consists of two charged ions, a positively charged sodium ion, Na^+, and a negatively charged chloride ion, Cl^-. Other 'salts', such as lithium and potassium chlorides, also elicit a salt response, though as you will discover the perceived response to these substances is more complex than for sodium chloride.

So, in contrast to the olfactory system, which detects volatile chemicals of small molecular mass, the gustatory system detects water-soluble materials of widely varying structure. An important consequence of this is that whereas the olfactory system can respond to distant stimuli, some originating possibly several hundred metres away, the gustatory system only detects stimuli that result from substances put directly into the mouth. The olfactory system, then, presents us with information about whether or not some odorous material is likely to be beneficial or harmful to us; the gustatory system provides us with the final choice as to whether or not we should ingest it. Perhaps this is why the gustatory system appears to have fewer perceived qualities compared to the apparently limitless numbers of odours we can detect. Once we put something into our mouths we need to make a very rapid decision about its desirability. Fewer choices make decision making easier.

How, then, do these molecules, which vary widely in molecular structure, bring about the perceptions they do? In the following sections we shall explore each taste quality in turn.

3.3 Summary of Sections 3–3.2

There appear to be five known taste qualities – bitter, sweet, umami, sour and salt – although these categories may be artificial constructs. These tastes are detected by taste cells housed in taste buds that can be found in the palate and in fungiform, foliate and circumvallate papillae on the tongue. In contrast to olfactory receptor cells, taste cells are specialized epithelial cells that synapse with neurons from either the chorda tympani (fungiform and anterior foliate papillae) or glossopharyngeal nerves (posterior foliate and circumvallate papillae). These two nerves go to the nucleus of the solitary tract. From here the pathway projects to the primary taste cortex, and thence to the amygdala and orbitofrontal cortex. The substances that elicit taste sensations have diverse molecular structures. They range in size from relative molecular masses less than 100 to masses greater than 20 000; some are ionic while others are covalent; and these substances are water soluble and generally involatile.

3.4 Bitter taste

Most people dislike bitter tastes and it is generally thought that our averse reaction to bitter tastes is an evolutionary one; many toxic substances taste bitter, so avoiding bitterness is a means by which we avoid ingesting poisonous materials. This may prove a correct hypothesis but as yet it is not firmly established. First, while it is true that some bitter substances, such as strychnine (Figure 3.7), are poisonous, the presence of others such as quinine and caffeine imparts desirable taste attributes to drinks. Quinine is the 'bitter' tastant in tonic water; caffeine is naturally present in coffee and tea and is commonly added to cola drinks, its enhancing effects often exerted at concentrations that are below its threshold of detection (about 0.7 mM or 140 mg l^{-1}). Beer also has many bitter constituents that are deemed to be desirable to the palatability of the drink – at least, they are to beer drinkers!

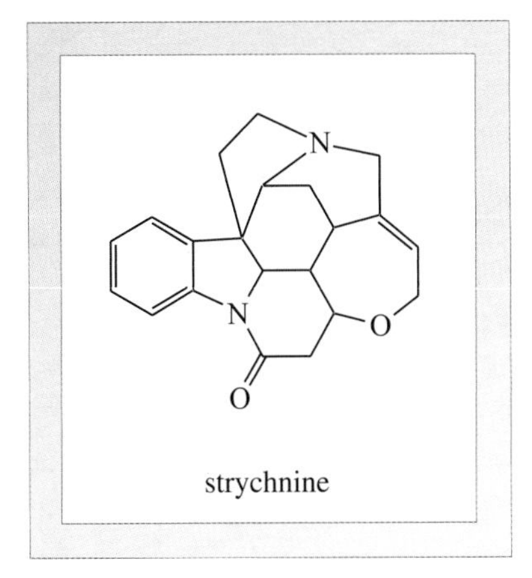

Figure 3.7 Strychnine, a poisonous bitter tasting substance.

Second, there appears to be a relationship between the general class of mammalian feeding type and bitter rejection thresholds to quinine (Table 3.1).

Table 3.1 Bitter taste threshold for quinine hydrochloride of various mammal feeding types.

Feeding type	Threshold
carnivores	0.02 mM
omnivores	0.30 mM
grazer herbivores	0.67 mM
browser herbivores	3.00 mM
humans	0.03 mM

❍ Which class of animal has the highest bitter rejection threshold and which has the lowest?

● Browser herbivores have the highest threshold and carnivores the lowest.

Since bitter substances are commonly of plant origin, animals that feed on such dietary sources are likely to develop, through evolutionary adaptation, some degree of tolerance for bitter taste. Equally, animals whose diet seldom includes bitter tasting materials – carnivores – are unlikely to have adapted to them and will exhibit a low tolerance for bitter tastants. Thus, unless each feeding type has concomitantly also developed specialized mechanisms for dealing with the toxicity of bitter substances it seems unlikely that equating bitter with toxic is proven.

Whatever the reason we have bitter taste cells, have them we undoubtedly do. However, the ability to perceive bitter tastes varies greatly across individuals. Indeed, of all the five taste modalities, bitter taste perception appears to be unique; it is the only one that has a recorded incidence of human **polymorphism** – that is, there are sub-sets of the human population that have different bitter taste perception profiles that are genetic in origin. Serendipitously, it was found that humans can be divided into two broad groupings, those that can taste the bitter substances N-phenylthiocarbamide (PTC) and 6-propylthiouracil (PROP) (Figure 3.8) and those that cannot. The ability to taste PROP is thought to be a dominant genetic trait, and in Caucasian populations 70 per cent are PROP tasters while in Asian and African-Americans an estimated 90 per cent are PROP tasters. It appears that the PROP tasters can be further subdivided into those who are medium tasters, for whom PROP is moderately bitter, and those who are 'supertasters', for whom PROP is intensely bitter. About one-third of PROP tasters are supertasters, most of whom are women. Supertasters tend to have more fungiform papillae – averaging 425 per cm^2 as compared with 184 and 96 per cm^2 for medium and non-tasters, respectively – and also a higher density of taste buds per papilla (Figure 3.9).

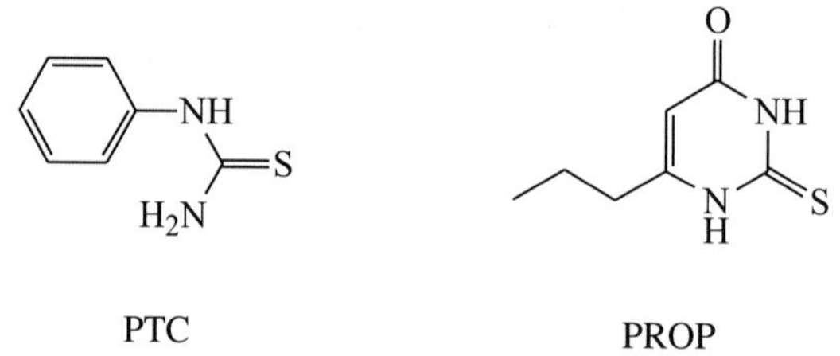

PTC

PROP

Figure 3.8 The bitter tasting substances PTC and PROP.

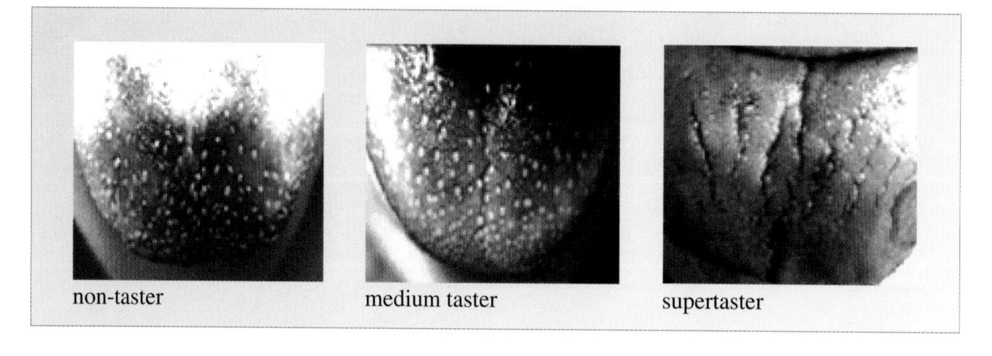

Figure 3.9 The tongues of three individuals of different PROP-taster status stained with blue food colouring. On the left is the tongue of a non-taster; the pink speckles are the fungiform taste papillae. In the middle, the tongue of a medium taster has a higher density of fungiform papillae. On the right, a supertaster's tongue is almost covered with fungiform papillae so that it is only lightly stained.

PROP tasters are reported to rate other bitter tastants – urea, denatonium benzoate, potassium chloride, quinine and caffeine – as more bitter than non-tasters. Since PROP taster status is correlated with fungiform papillae density, one might anticipate taster status to be associated with the perception of other tastants. Indeed, PROP tasters report sucrose as sweeter, salt as saltier, and citric acid as more sour. Such differences seem to be associated with different liking for tastes and with food preferences.

This difference in PROP taster status has proved invaluable in the search to uncover the molecular mechanism(s) of bitter taste. Given the diverse structures of bitter substances, it has generally been assumed that such molecules bind to a range of receptor proteins (rather than just one or two). Genetic studies have linked

PROP-taster status to a region of human chromosome 5. By inspecting the DNA sequences of this chromosome, the American scientists Charles Zuker and Nicholas Ryba identified a segment that coded for a protein that has an amino acid sequence typical of a G-protein-coupled receptor and which they believed to be a bitter taste receptor. They called this receptor T2R-1 (taste receptor family 2, member 1). By using the T2R-1 DNA sequence as a guide, Zuker and Ryba looked for related sequences and found twelve T2R genes, located on chromosomes 5, 6 and 12. However, the databases they used contained only a fraction of the human genome so they have estimated that humans have approximately 40–80 different functional bitter taste receptors.

Of course, finding the potential bitter taste receptor proteins is only the first step in identifying them as bitter taste receptors. The next step requires them to respond functionally to known bitter tastants. While this remains to be achieved for the human taste system, studies with rodents have established such a link. Both mice and rats – which have T2R proteins analogous to their human counterparts – express bitter receptor proteins in about 15 per cent of the cells in every taste bud of the circumvallate and foliate papillae, as well as in the taste buds in the soft palate. However, only about 10 per cent of fungiform papillae contain bitter taste receptors, though the density of bitter taste receptor cells in these is the same as elsewhere.

That T2Rs are bitter taste receptors can be seen when the bitter taste receptors are inserted into cells that do not normally respond to bitter tastants. Using a fluorescence method to measure intracellular Ca^{2+} changes (Sections 2.10 and 2.11), the cells respond rapidly and selectively to bitter tastants (Figure 3.10). Cells expressing the human T2R-4 (hT2R-4) respond strongly to denatonium benzoate, rather weakly to PROP and not at all to other bitter tastants.

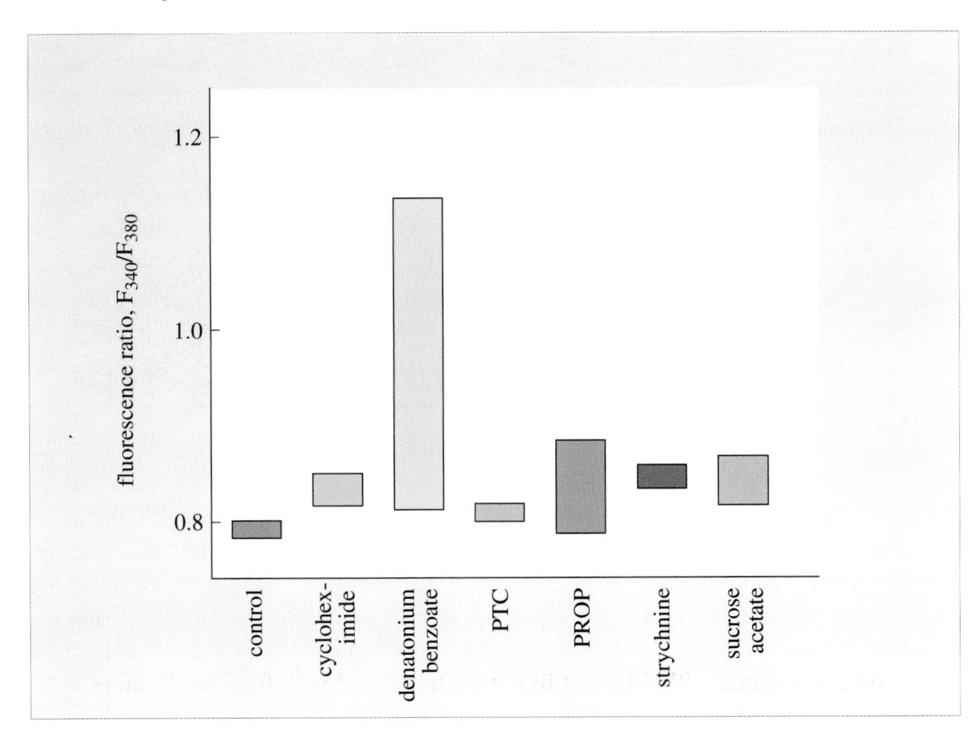

Figure 3.10 Fluorescence changes, brought about by several bitter tastants, observed in cells into which the human T2R-4 taste receptor has been inserted.

Box 3.1 Mouse polymorphisms for bitter taste

Mice have a taste receptor, mT2R-5, that responds to the bitter compound cycloheximide. Interestingly, mice exhibit a polymorphism for cycloheximide that is similar to the human ability to taste PROP. Cycloheximide-taster mice avoid solutions of cycloheximide that are ten-fold more dilute than non-taster mice. Significantly, mice that do not taste cycloheximide exhibit mutations of the mT2R-5 receptor protein at just five of the amino acid residues. The concentration of cycloheximide that can bring about half-maximal response of the T2R-5 receptor of taster mice – a concentration that reflects the strength of the binding of cycloheximide to the receptor – is 0.3 μM, whereas the corresponding value for the mutated T2R-5 receptor of non-taster mice is 2 μM. This nearly ten-fold difference in receptor sensitivity reflects the difference in cycloheximide concentrations that are avoided by taster and non-taster mice.

The human bitter taste receptors discovered so far contain about 300 amino acid residues. This makes them comparable in size to olfactory receptors. Like the olfactory receptors, the bitter taste T2Rs have a short extracellular N-terminal domain. While it is unclear how T2Rs recognize bitter tastants, comparison of the individual sequences of the proteins shows that transmembrane segments 1, 2, 3 and 7 have high similarity between different receptors and that structural diversity is highest in the extracellular loops extending partway into the transmembrane helices.

❍ Where does this suggest bitter tastants bind to T2Rs?

● To recognize structurally different tastants the T2Rs probably use the most variable segments of the receptor protein – that is the extracellular loops and possibly the portions of transmembrane helices nearest the extracellular surface. Alternatively, bitter tastants may bind to the transmembrane domains 4, 5 and 6.

Interestingly, for the mouse T2R-5 receptor that seems to be involved in responding to cycloheximide, the TM4–TM6 domains of tasters and non-tasters seem to have identical amino-acid sequences, whereas the differences in amino acid residues between the two are seen in the extracellular loops between TM2 and TM3, and TM4 and TM5 as well as near the extracellular surface of TM3.

One intriguing observation is that any one bitter taste cell appears to express many different bitter taste receptor proteins; indeed it may be that each bitter taste cell expresses all bitter receptor proteins! This suggests that the bitter taste cell can respond to a range of different bitter tasting molecules. This would be consistent with the human ability to perceive a general bitter quality rather than differentiate between differing bitter tastes. And yet, more recent studies have revealed that individual receptor cells can indeed discriminate between different bitter tastants. When bitter taste cell responses to five bitter tastants – cycloheximide, quinine, PTC, sucrose acetate and denatonium benzoate – were examined, 65 per cent of all cells responded to only one of the bitter compounds tested, 26 per cent responded to two bitter tastants and only 9 per cent to more than two (Figure 3.11). It may be, therefore, that a bitter taste perception is not the result of the activation of a homogeneous population of bitter-sensitive taste cells but via activation of multiple cells from a heterogeneous population.

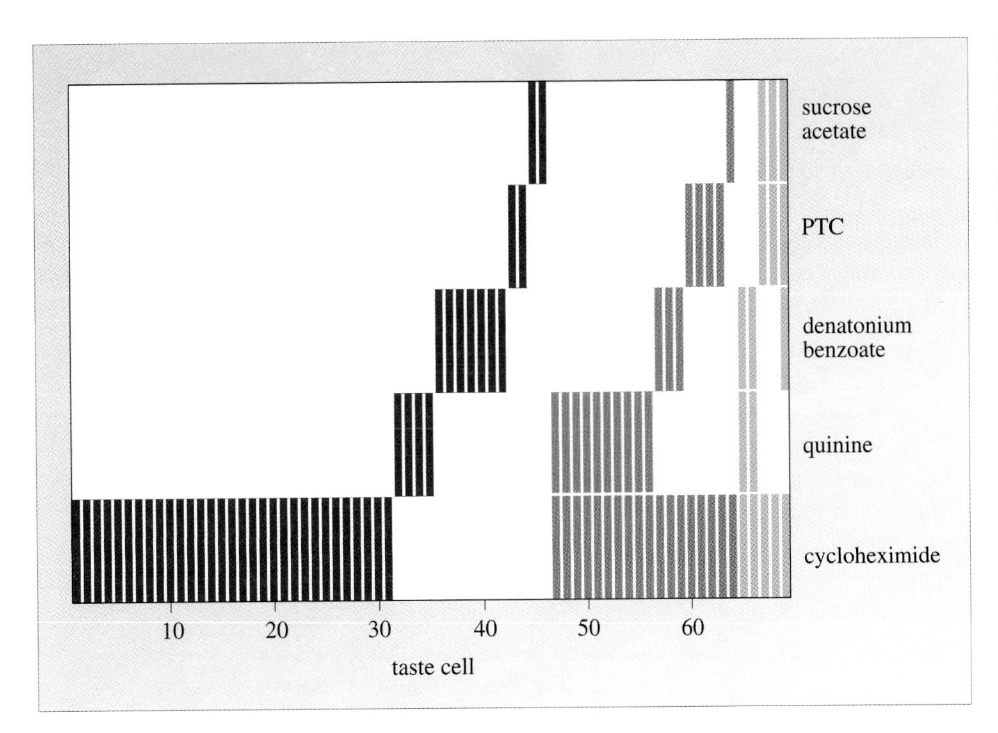

Figure 3.11 Response profiles of bitter-sensitive taste cells to five bitter tasting molecules. Maroon boxes represent cells responding only to one tastant, brown to two and pink to more than two.

❍ One of the above descriptions is consistent with a 'labelled-line' code, the other with a pattern code; which is which?

● The former, that one bitter taste cell expresses multiple receptors and produces a common response to exposure to bitter tastants, corresponds to the labelled-line concept. The latter, that different cells discriminate between bitter substances, is consistent with a pattern code.

Clearly there is much left to discover about how bitter tastants elicit a bitter taste!

3.5 Sweet taste

> Just a spoonful of sugar helps the medicine go down in a most delightful way.

So sang Mary Poppins in the Disney film of the same name. Indeed, in the UK most children used to be immunized against polio by ingesting the vaccine adsorbed on a sugar lump. Our ability to perceive sweet tastes, though, is unlikely to have developed to aid our ability to ingest otherwise unpleasant materials. More likely, it enables us to respond in a positive manner to foods with high calorific content. As Brillat-Savarin wrote:

> Taste…invites us, by arousing our pleasure, to repair the constant losses which we suffer through our physical existence.

The most common sweet substances that we encounter from the natural world are the carbohydrates fructose, glucose and sucrose (Figure 3.12 overleaf) and these provide ready sources of energy. On a gram for gram basis, sucrose and fructose are perceived as having approximately the same sweetness, both being about twice as sweet as glucose. However, the relative molecular mass of a sucrose molecule, $M_r = 342$, is nearly twice that of glucose and fructose, both $M_r = 180$, so on a

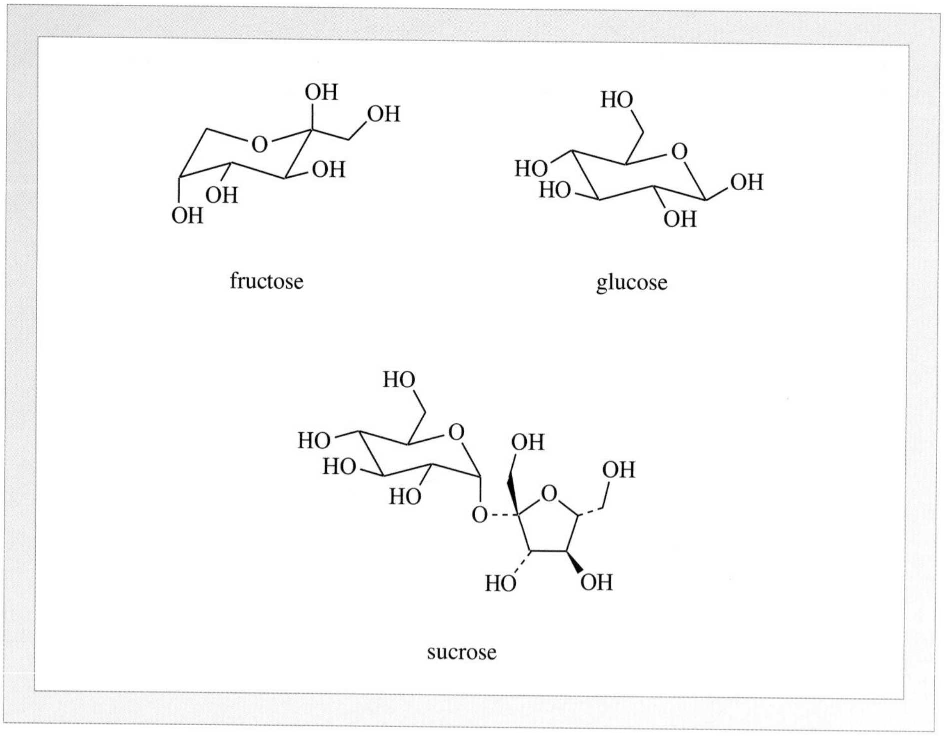

Figure 3.12 The molecular structures of fructose, glucose and sucrose.

molecule for molecule basis sucrose is twice as sweet as fructose and four times as sweet as glucose. How do these molecules produce these different taste perceptions?

Until very recently, the only molecular leads that taste scientists had to go on were the structures of sweet-tasting molecules themselves. Alongside the sweet-tasting carbohydrates, a range of other sweet-tasting molecules is known. Some we have mentioned already, such as saccharin and glycerol, and others, such as aspartame, sodium cyclamate and acesulfame K (Figure 3.13), you may recognize from the ingredients labels of food packaging.

These molecules don't appear to have much in the way of molecular similarity – only the carbohydrates seem to have anything in common – but by comparing the various features in all of these molecules, together with those of other sweet tasting substances, taste scientists recognized that it was almost certain that they were all eliciting a sweet taste through binding to a receptor protein housed in taste cells.

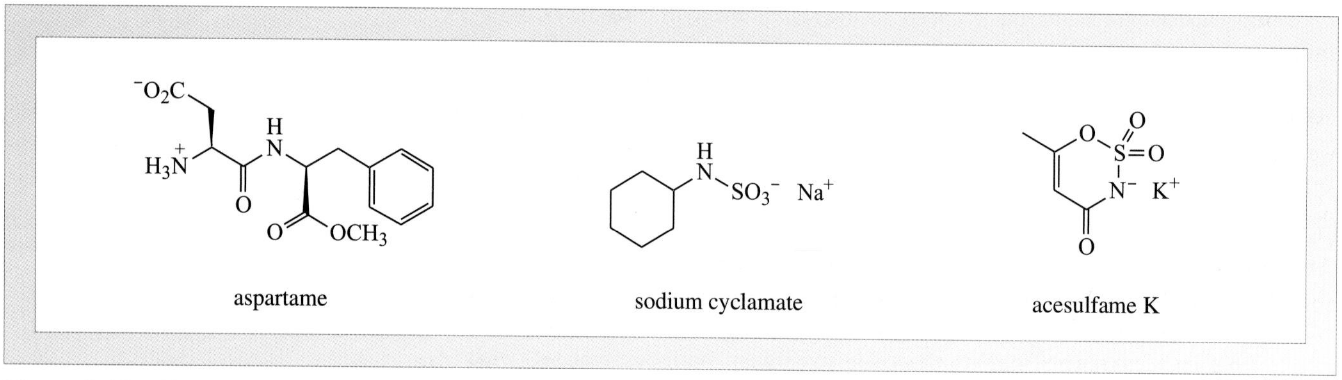

Figure 3.13 The molecular structures of aspartame, sodium cyclamate and acesulfame K.

Though the nature of this sweet receptor remained unknown, the structures of the sweet tastants were used to construct various models of the receptor binding interactions. Several have been put forward, varying in their degree of complexity (Figure 3.14).

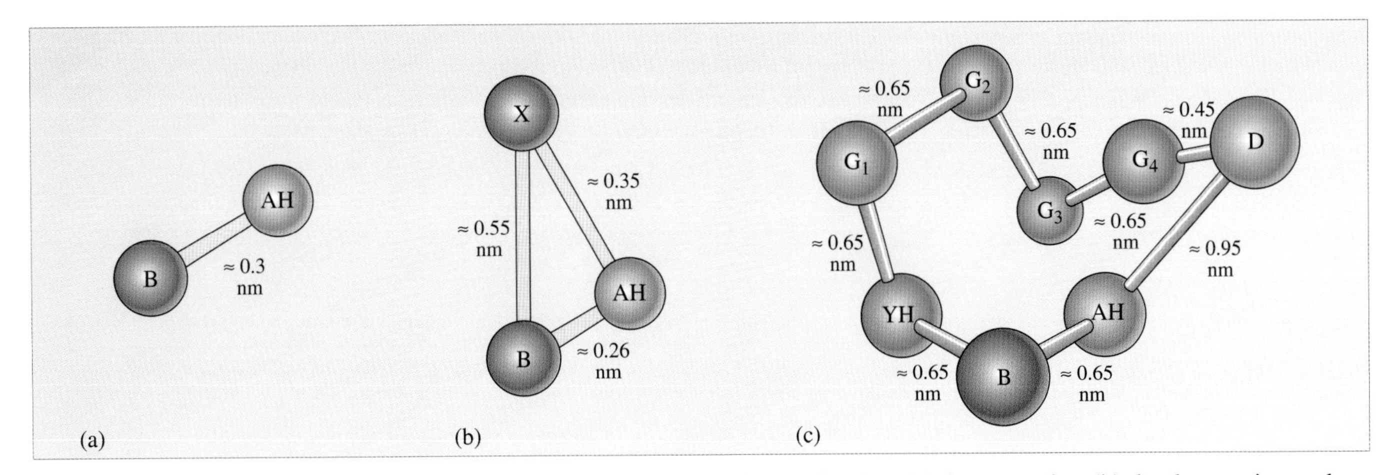

Figure 3.14 Three models of receptor interactions with sweet tasting molecules: (a) the two-point, (b) the three-point, and (c) the multi-point attachment models.

The simplest, first suggested in 1967 by American scientists R. S. Shallenberger and T. E. Acree, is a two-point attachment mechanism involving hydrogen bonding with the receptor of two groups on the sweet tastant situated 0.25–0.4 nm apart. One of these groups, termed AH, donates a hydrogen bond, the other, B, accepts a hydrogen bond.

❍ Suggest a potential weakness of such a mechanism.

● Many molecules are likely to have such an arrangement, so it would imply a large number of sweet-tasting molecules, which seems unlikely.

A modification was proposed by L. B. Kier in 1972, requiring sweet-tasting molecules to possess a third component for receptor interaction. Labelled X in Figure 3.14b, this is a lipophilic region, positioned about 0.35 nm from AH and 0.55 nm from B, that interacts with a similar region on the receptor.

Many of the synthetic sweeteners with which we come into contact in everyday life have been discovered using these or similar models for tastant-receptor binding. Indeed, using these models, molecules that are many times sweeter than the natural carbohydrate sweeteners have been designed and synthesised. For example, aspartame is 180 times as sweet as sucrose. In terms of weight control in the modern world, sweeteners such as these have the advantage over carbohydrates in that smaller quantities are required to achieve a similar intensity of taste perception, so they are less calorific.

As further examples of synthetic sweet-tasting molecules have been discovered the complexity of the models has increased. The most complex, using all known sweet tastants, is the multipoint attachment theory of the French scientist Claude Nofre, originally proposed in 1991 and updated in 1996, in which *eight* interaction sites are considered important, each of which can involve both hydrogen bonding and some kind of steric or lipophilic interaction with the receptor! Sweetness potency is

Box 3.2 Measuring sweetness

Sweetness is usually measured by comparison with reference solutions of sucrose. Humans can perceive a 2 per cent sucrose solution as sweet and a 15 per cent sucrose solution as very sweet. The sucrose concentration in a typical cup of sweetened coffee is about 5 per cent. The intensity of a sweet taste is usually quantified by means of a scale line 15 cm long, using 2 to 15 per cent solutions of sucrose as references (Figure 3.15).

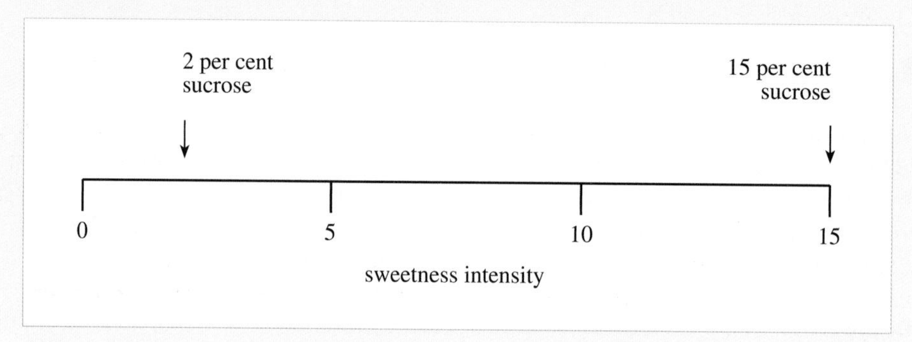

Figure 3.15 A scale used for measuring sweetness intensity.

thought to be related to the number of interactions between the tastant and the receptor; the greater the number of the interactions the more potent the sweetener.

Although such models have proved useful for the ongoing discovery of novel synthetic sweeteners, it does not mean they correctly represent the potential tastant-receptor interactions for any particular sweet tastant. The following questions perhaps clarify this issue.

❍ What assumption is made about the number of sweet taste receptors that exist when the model of tastant-receptor interactions is developed using the structure of all tastants?

● This method develops a model of binding that assumes the existence of only one type of sweet taste receptor.

❍ From what you have discovered about bitter taste receptors, do you think a universal sweet taste receptor to be likely?

● It is highly unlikely. Up to eighty bitter taste receptors may exist, so in all probability there is more than one sweet taste receptor.

However, one might anticipate that, in contrast to bitter taste, if the number of biologically sweet tastants is limited (evolution having not yet responded to synthetic sweeteners!) the number of sweet receptors may be small. Indeed, this looks to be the case.

Following on rapidly from the discovery of bitter taste receptors in 2000, by mid-2001 the existence of functional sweet taste receptors was reported separately by the groups of Ryba and Zuker on the one hand and Buck on the other. The discovery of the sweet taste receptors was aided by the existence of two strains of mice, one that is able to taste the synthetic sweetener saccharin and one that is a saccharin non-taster. The difference between these two strains was shown to be genetic, and was mapped to mouse chromosome 4. By inspecting the mouse genome it proved possible to identify the code for a G-coupled-protein receptor. Significantly, this receptor, called T1R-3, when inserted into non-taster mice enabled these mice to recover their ability to taste saccharin but did not affect their ability to respond to other types of taste such as bitter (cycloheximide), salt (sodium chloride), sour (HCl) and umami (monosodium glutamate).

❍ What does this imply about the role of T1R-3?

● It implies that it functions as a sweet taste receptor, at least for saccharin.

Of significance to the human perception of taste, this mouse sweet taste receptor has a human counterpart, the code for which is located on human chromosome 1. From the amino acid sequences, 72 per cent of the human receptor is identical to its mouse counterpart. Both are about 820 amino acid residues long, giving them a relative molecular mass of about 125 000 (enormous compared to the relative molecular mass of 180 for fructose), the first 530 or so forming a large extracellular domain in the region of the N-terminus (Figure 3.16 overleaf).

Despite the large size of this domain, it appears that only a few of the amino acid residues are crucial to tastant binding, at least for saccharin. For example, the T1R-3 receptor proteins of saccharin taster and non-taster mice differ in only six of their residues; four of these are in the N-terminal region, one lies near the extracellular side of the fourth transmembrane domain (TM4), and the sixth in the intracellular C-terminal domain.

❍ Where does this imply that saccharin binds to the receptor, and how does this contrast with the binding of bitter tastants to the bitter receptors?

● Given that the differences lie in the extracellular N-terminal domain it implies that sweet tastants bind here rather than, as is likely for bitter tastants, in the extracellular loops connecting the transmembrane spanning regions or in the transmembrane domains themselves.

However, it isn't yet clear whether a tastant like saccharin binds to a single receptor protein or whether it binds to two neighbouring proteins. That is because of an interesting feature of the pattern of expression of the T1R-3 receptor protein. It turns out that about 15–30 per cent of all cells from taste buds in the circumvallate, foliate and fungiform papillae, and a similar number of palate taste buds, express the T1R-3 receptor protein. This pattern of expression is similar to those for two other taste receptor proteins of the same T1 family, T1R-1 and T1R-2, for which no known role had been discovered. When the distribution of cells expressing these different receptors is compared the following patterns emerge: cells that express T1R-2 always express T1R-3; cells that express T1R-1 also express T1R-3; and some cells only express T1R-3. Based on this pattern of expression, there appear to be three known different types of sweet taste cells. So far, only those cells expressing both T1R-2 and T1R-3 have been examined in any detail. Using fluorescence imaging, the changes in intracellular Ca^{2+} were monitored in response to a range of sweet tastants (Figure 3.17 overleaf).

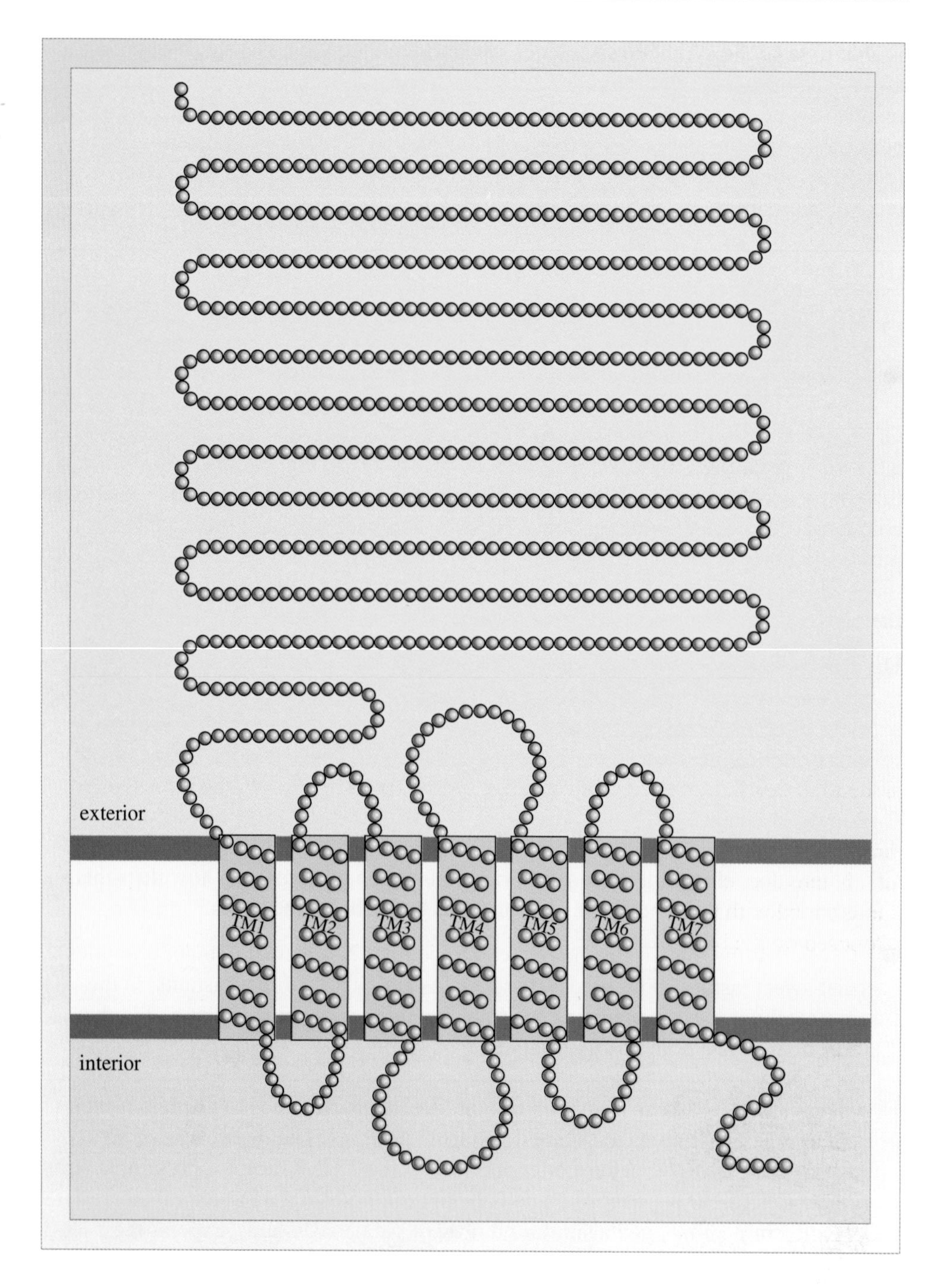

Figure 3.16 Schematic representation of the mammalian sweet taste receptor. Note the relative extent of the extracellular domain.

- ❍ Assuming that the responses are brought about by interaction of the sweet tastants with the receptors, what do the data in Figure 3.17 tell you about the general nature of such interactions?

- ● The interactions are selective. Some tastants elicit a strong response whereas others essentially none. Moreover, of those that do elicit a response some do so at much lower concentrations than others. For example the strongest response is to acesulfame-K (ace-K), which is present at 10 mM, whereas the fructose response is less than half that of ace-K but its concentration (250 mM) is 25 times that of ace-K.

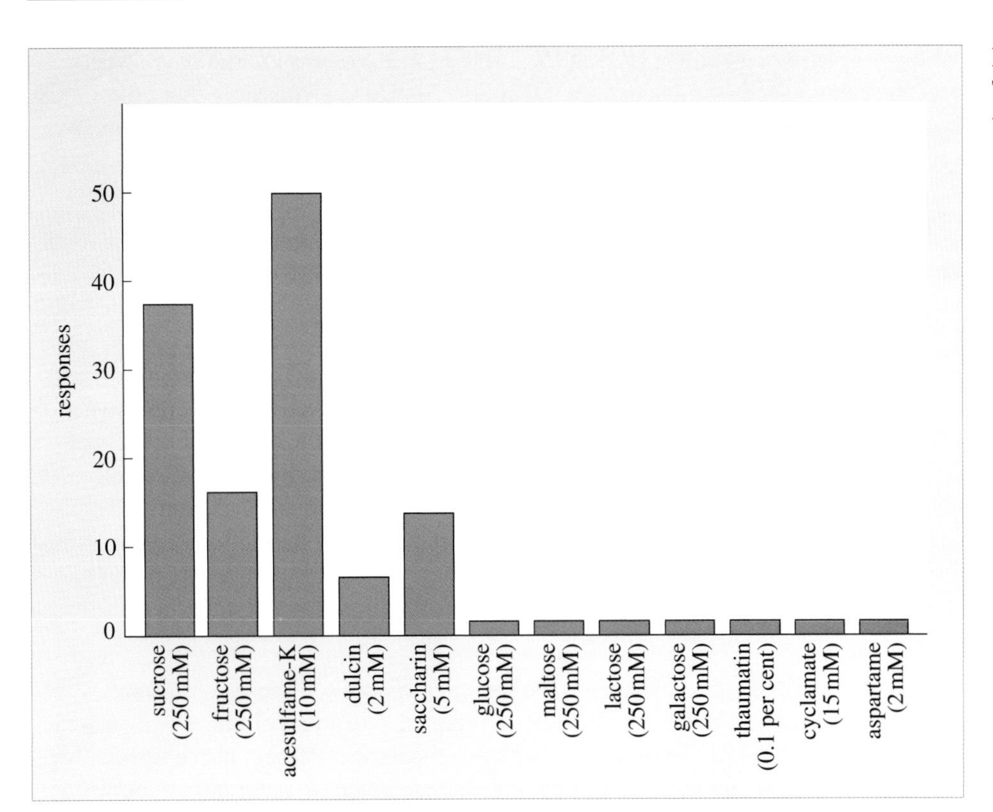

Figure 3.17 Responses of the T1R-2/T1R-3 receptor combination to several sweet tastants.

Crucial to the response to these sweet tastants is the presence of *both* T1R-2 and T1R-3; lack of either destroys the response. This has led to a proposal that the two proteins combine to form a structure that functions as the receptor (Figure 3.18). If that is the case, binding between sweet tastants and receptor proteins is somewhat more complex than for bitter tastants and even more complex than was originally proposed on the basis of the structures of sweet tastants.

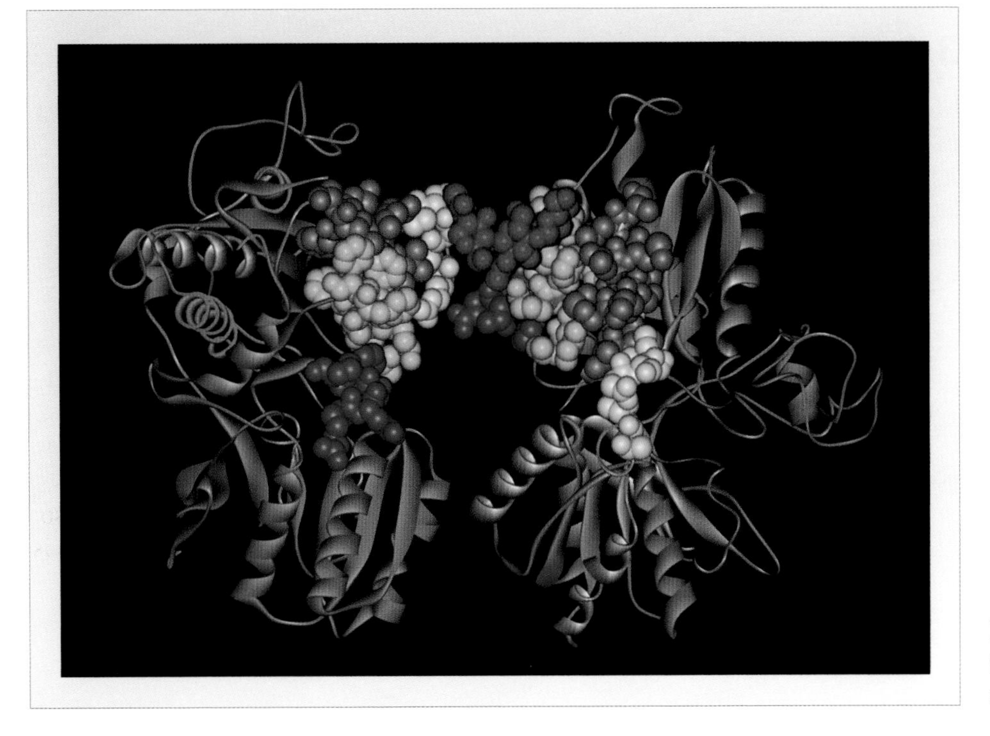

Figure 3.18 The dimeric sweet taste receptor formed from T1R-2 and T1R-3 receptor proteins.

At the time of writing, the T1R-1, T1R-2 and T1R-3 proteins are the only sweet receptors that have been discovered. Of course, it is always possible that others exist but even with these three there are nine possible combinations that could form sweet receptors, the three single receptors and six dimeric structures.

❍ Write down the possible combinations.

● T1R-1, T1R-2, T1R-3, T1R-1/T1R-1, T1R-1/T1R-2, T1R-1/T1R-3, T1R-2/T1R-2, T1R-2/T1R-3, T1R-3/T1R-3.

Not all of these may be functional. For example, T1R-1 and T1R-2 proteins do not seem to be expressed by the same cell to any extent. But even if up to eight sweet taste receptors do exist, how can we explain the ability of the chemical models of tastant–receptor interactions to predict structures of sweet tastants? In all likelihood, each receptor uses some of the different predicted interactions. Thus, some molecules bind to one type of receptor using a particular combination of interactions and others to a second receptor using a different combination of interactions. The three-point attachment theory probably contains the most basic requirements for these tastant–receptor interactions.

Like any chemosensory mechanism, the interaction between the sweet tastant and receptor protein(s) is the first event in a sequence of intracellular events which results in the release of the neurotransmitter that fires the afferent nerve fibres. The cell depolarization that triggers neurotransmitter release is brought about by a range of second messengers: some cells appear to use cyclic AMP, others cyclic GMP and yet others inositol-1,4,5-triphosphate. These signalling pathways are not yet clearly understood. However, the distribution of sweet receptors in the oral cavity – the T1R-2/T1R-3 cells at the back of the tongue and palate, the T1R-1/T1R-3 cells at the front – suggest that the cells expressing different combinations of receptor proteins will be differently innervated and will therefore exhibit different neural pathways.

3.6 Umami taste

Most people are familiar with four taste categories: sweet, salt, bitter and sour. Most of us are also aware of a taste quality we might define as 'savoury', although we might find it difficult to describe it well. Indeed, it is more than likely we would attribute such a quality to combinations of the more 'common' tastes, possibly supplemented by complex aromas. To achieve this savoury effect, we often prepare foods by adding sauces that impart a characteristic quality during cooking. In the West, these are often stocks, concentrated protein or yeast extracts (e.g. Bovril and Marmite) or tomato concentrates; in the East, fish and soy sauces have a similar role. What these all have in common is that they are derived from protein-rich food sources and contain high levels of 'free' amino acids, particularly glutamic acid. (Proteins are made up from amino acids covalently linked together; when the sauces are prepared these covalent links are broken to liberate the free amino acids.) Back in 1912, the Japanese scientist Kikunae Ikeda suggested that:

> ... an attentive taster will find something common in the complicated taste of asparagus, tomatoes, cheese and meat, which is quite peculiar and cannot be classified under any of the [other known tastes].

Ikeda had noticed that a broth made from the seaweed kombu had this distinctive taste and he set about trying to identify the component(s) responsible for it, eventually isolating the amino acid, glutamic acid (Figure 3.19) (100 g of dried kombu yields about 1 g of glutamic acid). This compound has a distinctive taste, different from sweet, salt, bitter and sour, and it is this distinctive taste that Ikeda called *umami*. Glutamic acid is a natural constituent of many foods, including fish, meats, cheese, mushrooms and some vegetables (Table 3.2).

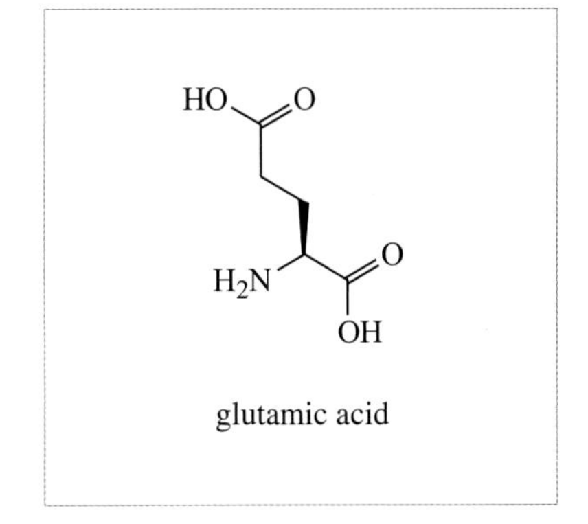

Figure 3.19 The structure of glutamic acid.

Table 3.2 Glutamate content of selected foods.

Food	Free glutamate (kg^{-1})
cod	90 mg
peas	2 g
tomatoes	1.4 g
carrots	330 mg
milk	20 mg
cheese	12 g
eggs	230 mg
beef	330 mg

Ikeda went on to develop the well-known food additive monosodium glutamate (MSG) as a culinary form of glutamic acid (glutamic acid contains two carboxylic acid groups; turning one of these into its sodium salt forms MSG).

Given that the umami taste, and a compound that could elicit it, was discovered almost one hundred years ago, it seems remarkable that the existence of a fifth taste type was disputed until very recently.

❍ As MSG contains both a carboxylic acid group and a sodium carboxylate suggest two other taste qualities it might elicit.

● The acid functionality might well elicit a 'sour' taste, and the sodium salt a 'salty' taste.

Consequently, it could be argued that umami is a particular combination of these two tastes. However, when presented to a panel of sensory tasters, umami was identified as being distinct from the other four tastes as well as from mixtures of them. That umami has an effect different from the sodium ion has been demonstrated using the diuretic compound amiloride (Figure 3.20). Amiloride is able to block Na^+ transport through the Na^+ ion channels of the apical epithelial membrane, thereby inhibiting the response of the chorda tympani (cranial nerve VII) to NaCl. The response to MSG in the presence and absence of amiloride revealed that the chorda tympani responds even in the presence of amiloride, implying that the umami taste sensation is physiologically distinct from that of NaCl.

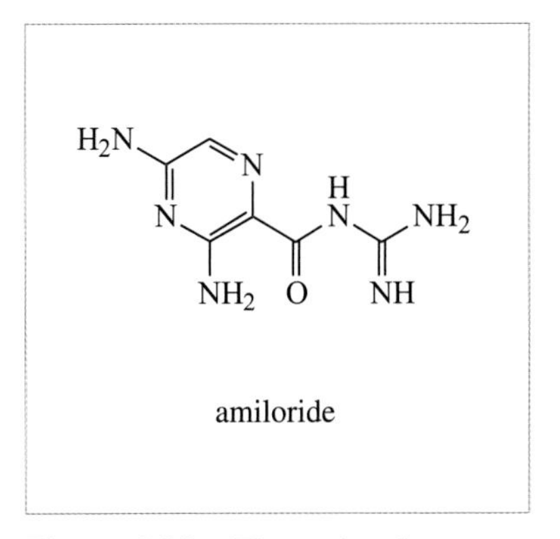

Figure 3.20 The molecular structure of amiloride.

Any doubt that umami is a distinct taste was removed in 2000 when the American scientist Nirupa Chaudhari demonstrated the existence of a glutamate taste receptor.

There is something of an irony here, because the umami receptor was the first of the three receptor-mediated taste mechanisms – sweet, bitter, umami – to be reported! Glutamate is a well-known neurotransmitter, with a multiplicity of receptor-binding sites, so the concept that it could bind to a glutamate receptor in taste cells is not a difficult one to comprehend. However, the receptors in the neurotransmitter system respond to glutamate concentrations in the *micromolar* (μM) range whereas glutamate concentrations in food are in the *millimolar* (mM) range.

- ❍ If the umami taste receptor were simply a glutamate neurotransmitter receptor and you compare the above two concentration ranges, what conclusion do you come to about the activity of an umami taste receptor?
- ● Glutamate in food is about one thousand times more concentrated than is required to activate a glutamate neurotransmitter receptor. If such a receptor functioned as an umami receptor in a taste cell it would function at maximum activity whenever it was exposed to glutamate-containing foods.

For that reason, Chaudhari and her co-workers set about looking for a receptor that was expressed only in taste cells and that responded to glutamate in the millimolar range. Using molecular biology techniques, they identified such a glutamate receptor from taste cells in circumvallate and foliate papillae. Like bitter and sweet receptors, this is a G-protein-coupled receptor that is membrane bound and contains seven transmembrane-spanning helical domains. It comprises about 475 amino acid residues and has a relative molecular mass of 68 000.

- ❍ Is the taste glutamate receptor larger or smaller than the bitter and sweet receptors?
- ● Bitter receptors comprise about 300 amino acid residues, so the glutamate taste receptor is about 50 per cent larger, while sweet receptors contain about 830 amino acid residues making them twice as large as the umami receptor.

These differences are reflected in the lengths of the extracellular domains in the region of the N-terminus (Figure 3.21). In bitter receptors this N-terminal chain is short, in the glutamate taste receptor it is much longer, while that in the sweet receptors is longer than the entire umami protein itself. Significantly, this N-terminal region is where the glutamate taste receptors and the glutamate neurotransmitter receptors differ, with the N-terminus of the glutamate neurotransmitter receptor being much longer.

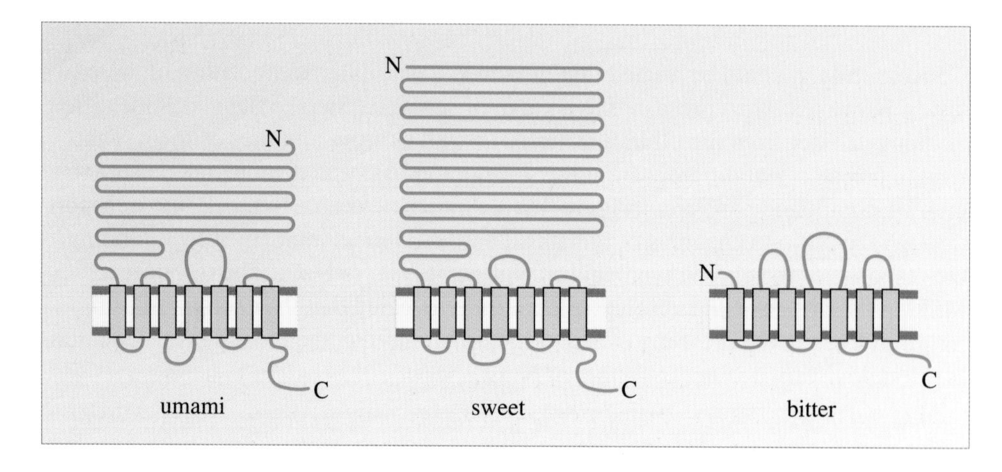

Figure 3.21 Probable membrane topologies of umami, sweet and bitter receptors.

This glutamate taste receptor protein responds to glutamate in the millimolar range, the affinity being in the same range as the taste threshold for umami, which is about 2 mM (300 mg l^{-1}). This reduced sensitivity to glutamate (compared to the corresponding neurotransmitter receptors) is thought to be due to the truncated extracellular N-terminal domain, and it is here that glutamate is thought to bind to the receptor protein.

Although glutamate is the prototypical compound eliciting an umami taste, it is not the only such compound. As you might expect, compounds with similar molecular structure also elicit such a taste sensation. Monosodium aspartate (MSA), the sodium salt of aspartic acid (Figure 3.22) is one such compound, although the detection threshold for MSA is 7.5 mM (1 g l^{-1}).

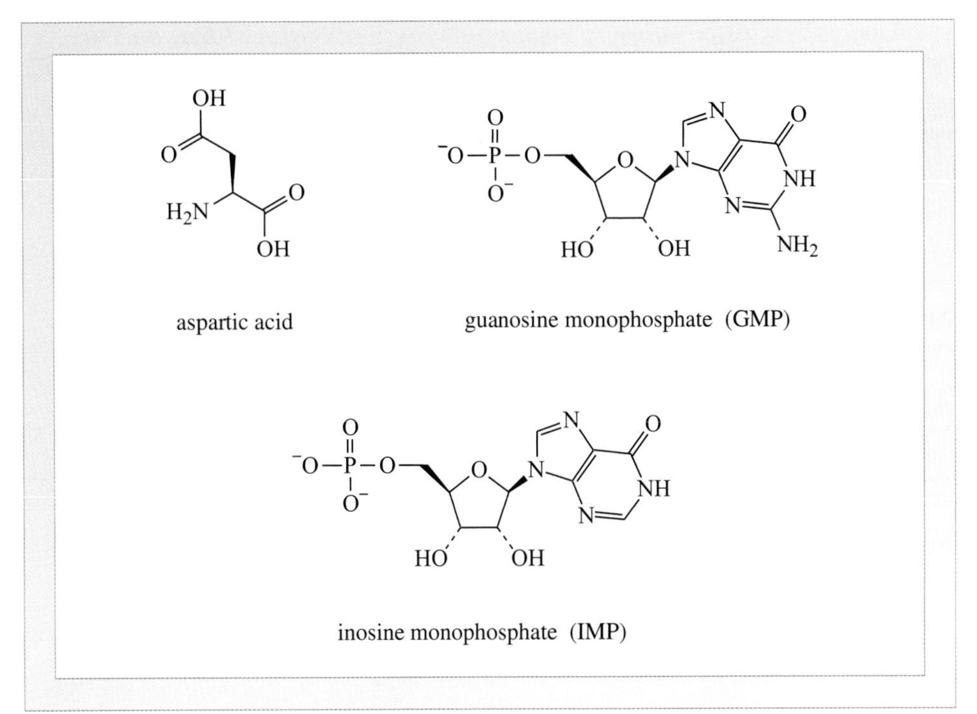

Figure 3.22 Some compounds that elicit an umami taste.

❍ Suggest a plausible rationale for the difference in threshold detection between MSG and MSA.

● MSG evokes a response at lower concentration. In all likelihood, MSG binds to the glutamate taste receptor with a higher affinity than does MSA. Presumably the slight difference in molecular structure means that the groups that interact with the receptor are better positioned in MSG than in MSA.

Compounds of rather different structure, such as the ribonucleotides GMP and IMP, also elicit an umami taste. It is unlikely, given their different molecular structures, that they bind to the umami taste receptors in the same way as glutamate. Currently, it is unknown how these compounds elicit an umami taste. However, there is a strong taste synergism between glutamate and the ribonucleotides; the umami taste intensity of a solution containing MSG and ribonucleotide is perceived as being stronger than the sum of the individual perceived intensities of glutamate and ribonucleotide. Additionally, the presence of a ribonucleotide lowers the taste threshold of MSG, and vice versa; for example, IMP lowers the MSG taste threshold by more than fifty-fold.

❍ Are the above observations consistent with MSG and ribonucleotides binding to the same receptor site of the glutamate taste receptor?

● No, they are not. If they bind to the same site they must compete with each other for that site. Consequently, the perceived intensity of the mixture would not be greater than that of the component that brought about the stronger individual taste intensity.

These observations seem to point to an alternative mechanism of action. One possibility is that glutamate binds to a different part of the glutamate receptor protein such that when glutamate is bound it affects how the ribonucleotide binds, and vice versa. The effect of the two bound together is therefore different from those of the two bound separately, and such an effect could be greater than the sum of the individual effects. Alternatively, glutamate and the ribonucleotides may bind to separate receptor proteins on the same umami-sensing taste cell. A third possibility is that they bind to two separate gustatory cells that converge onto the same afferent nerve fibres.

Clearly, our understanding of umami taste is in its infancy. We still do not know how the brain determines the taste of an amino acid, although it seems likely that, from an evolutionary point of view, the umami taste guides us to protein-rich food sources. Indeed, when MSG is added to foodstuffs at an optimal concentration of 0.6 per cent, it has been found that the intake of 'savoury' foods is increased.

3.7 Summary of Sections 3.4–3.6

Bitter, sweet and umami taste qualities are tranduced by taste cells through the involvement of membrane-bound G-protein-coupled receptors.

Bitter taste quality may be an averse response to limit exposure to toxic materials. Humans are polymorphic in their response to the bitter stimulus PROP: some are supertasters, some are medium tasters and some are non-tasters. Bitter tastants are structurally diverse and are recognized by the T2R family of receptors, of which there are likely to be up to 80 members. These receptors are about 300 amino acid residues in length, the N-terminus being relatively short. Tastants are thought to bind to the extracellular loops linking the transmembrane domains.

The sweet taste quality is probably a means of recognizing high calorific foods containing carbohydrates. Synthetic sweeteners are more potent and therefore less calorific. Sweet tastants are also structurally diverse and are recognized by a family of three T1R receptors. These are large proteins of about 830 amino acids with an extremely long N-terminal domain. At least one dimeric receptor, formed from T1R-2 and T1R-3, is known, but it is not yet clear how sweet tastants interact with the receptor proteins.

The umami taste quality appears to be involved in recognizing food sources high in protein. Umami tastants include glutamic acid and ribonucleotide monophosphates. These bind to a 475 amino acid residue umami receptor that has a medium sized N-terminal domain. The effects of glutamic acid and ribonucleotide monophosphates are synergistic.

3.8 Sour taste

The English word 'sour' derives from the same source as the German 'sauer' (as in sauerkraut). In German, the word is used to name a particular class of chemical compounds, the acids. So for example, buttersäure (booter-soy-ruh) (Figure 3.23) is the German for butanoic acid, an acid derived from butter (the literal translation is 'butter acid'), and citronensäure (see-troanen-soy-ruh) is the name for citric acid (literally 'lemon acid').

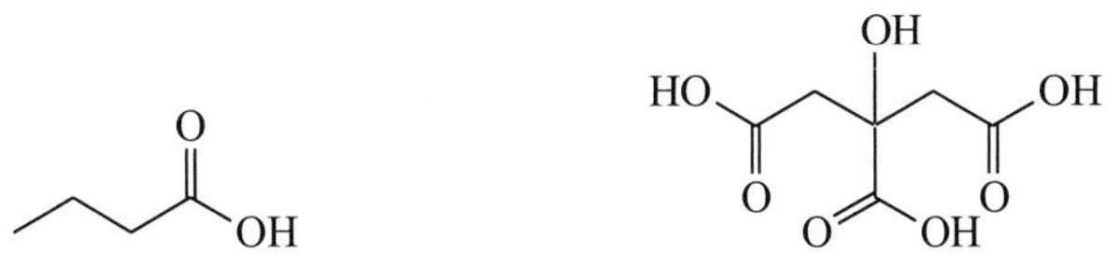

Figure 3.23 The molecular structures of butanoic acid and citric acid.

In German, then, the connection between sour taste and acidic compounds is explicit in the language, something that is not quite so obvious in English. Even so, it is generally recognized that acidic materials give rise to sour tastes. In fact, it is the case that the sour response is brought about uniquely by acids.

It is usually considered that sourness, like bitterness, is an averse reaction that protects us from ingesting harmful or unpleasant foods. Certainly, infants display an innate rejection response to it, and the averse sensation generally remains throughout life. Often, acids are produced as part of the biochemical processes of decay – acetic acid when wine is over-fermented, and butanoic acid when butter turns rancid. Unripe food also contains significant quantities of acidic materials. However, we should be wary of overstating this connection. Acetic acid, generally encountered in the home as vinegar, is used to pickle foods and thereby preserve them, and is also used as a condiment to flavour foods. So the sour taste of acetic acid is not entirely perceived as unpleasant. Moreover, ascorbic acid is a sour-tasting substance that is essential to human health – its more common name is vitamin C.

Though acids from foodstuff sources – like citric acid, ascorbic acid or vinegar – elicit a sour taste perception, it is not only organic acids (acids derived from compounds of carbon) that taste sour. Inorganic, or mineral, acids – hydrochloric acid (HCl), sulfuric acid (H_2SO_4) and the like – are also perceived as sour. The one feature that all acids, both organic and mineral, have in common is that in aqueous solution they give rise to hydrogen ions, H^+. This similarity can be seen from the following equations:

organic acids: $RCO_2H = H^+ + RCO_2^-$

inorganic acids: $HX = H^+ + X^-$

where RCO_2H is the general formula for an organic acid (compare the general formula with the structures in Figure 3.23) and HX is the general formula for an inorganic acid (for example, for hydrochloric acid X is Cl). So it would appear that the taste receptor cells that respond to acids do so by responding to H^+ ions. But how?

❍ Do you think sour taste receptor cells respond by a mechanism that involves binding of an acid molecule to a receptor protein?

● It seems unlikely, because the taste response is not related to the overall molecular structure of the acid, only to H^+.

Indeed, it is very unlikely that sour taste is brought about by a receptor binding mechanism. The H^+ ion is very small and it is spherically symmetrical. The latter means there is no directional preference for any molecular interactions with a receptor protein. As you discovered in Sections 2.2 and 2.3, receptor binding usually involves multi-atom molecules that have distinct directional preferences in their interactions with receptor proteins.

An alternative possibility could be that the sour taste receptor cells simply respond to hydrogen ion concentration in the vicinity of the cell. In other words, the sour taste receptor cells function as pH detectors of the external environment. A simple test of such a hypothesis would be to measure the sour response to two different acids applied at equal hydrogen ion concentrations (i.e. equal pH values). However, when hydrochloric and acetic acids are tested in this way, acetic acid is perceived as being much the more sour. This enhanced perception of acetic acid appears to have a physical basis. Figure 3.24a shows the electrical activity of a rat left chorda tympani when the anterior surface of the tongue was exposed to solutions of different acids at the same pH. Quite clearly, the response is greater to acetic acid than to hydrochloric acid, an indication that under these conditions acetic acid effects the stronger sour-taste stimulus.

One important point to note here though is that whereas the pH 3 hydrochloric acid solution has a total acid concentration of 10^{-3} mol l^{-1}, the pH 3 acetic acid solution has a much larger *total* acid concentration of approximately 10^{-2} mol l^{-1} (see Box 3.3 overleaf). Another curious observation, at least if our sour taste receptor cells function as pH detectors, is that the perceived sourness of two acetic acid solutions that differ in their pHs but contain equal total amounts of acid (the sum of the CH_3CO_2H and H^+ concentrations) is the same (see Box 3.3). Again, this appears to have a physical basis. The responses of the chorda tympani to taste cells exposed to acetic acid solutions of different pH, but of the same total acid concentration, are shown in Figure 3.25.

❍ Compare the following traces in Figure 3.25: (a) with (b), (c) with (d) and (b) with (d). What can be said about the nerve response to (i) solutions of the same pH but differing amounts of total acid, and (ii) solutions of differing pH but equal amounts of total acid?

● Comparing (b) with (d), the traces of two solutions of equal pH, it is clear that the solution of greater total acid produces a larger nerve response. In contrast, comparison of (a) with (b) shows that the nerve response is identical in both despite the two solutions having different pH values. A similar observation can be made from traces (c) and (d). It would appear that nerve response is directly related to the concentration of total acid present.

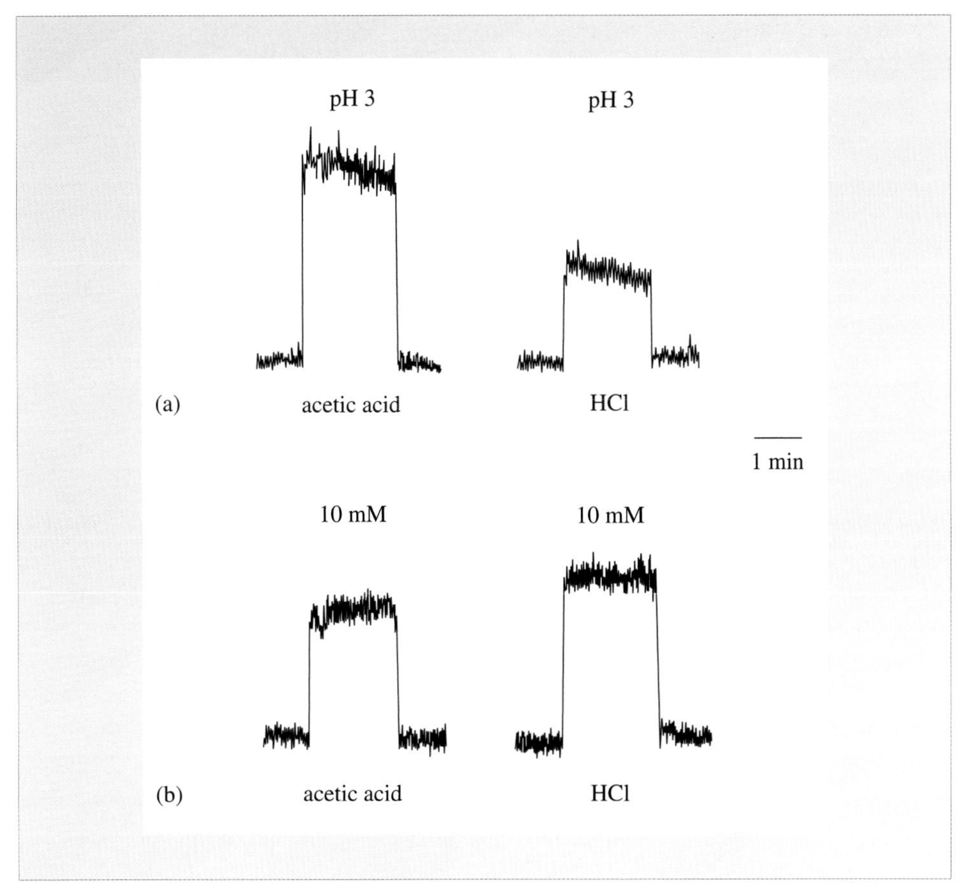

Figure 3.24 Chorda tympani activity resulting from two acid stimuli (a) at constant pH, and (b) at constant concentration.

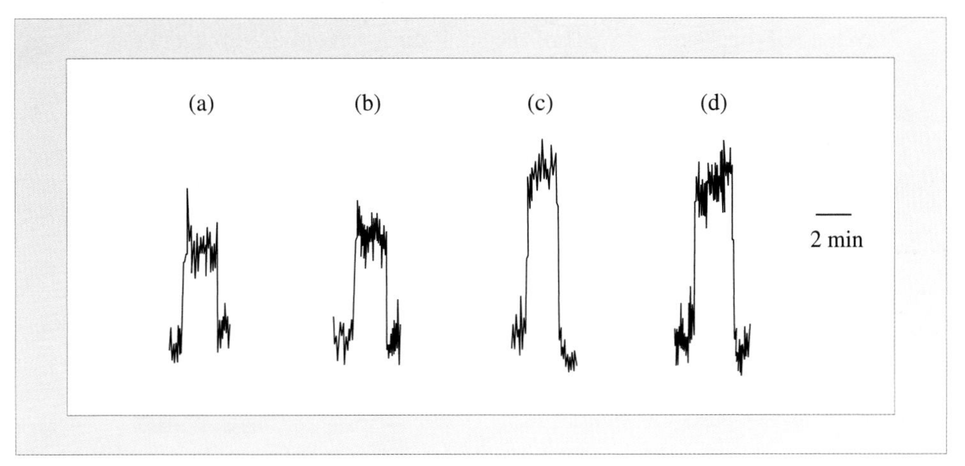

Figure 3.25 Chorda tympani responses to (a) 10 mM acetic acid (pH 3.4), (b) 10 mM acetic acid + 175 mM potassium acetate (pH 6), (c) 30 mM acetic acid (pH 3.15), and (d) 30 mM acetic acid + 525 mM potassium acetate (pH 6).

Box 3.3 pH and aqueous solutions of acids

When mineral acids dissolve in water the HX molecules dissociate entirely into the ions H^+ and X^-. So, all of the HX put into solution exists as ions and the concentration of H^+ is the same as the initial concentration of HX.

In contrast, when most organic acids dissolve in water the RCO_2H molecules do not dissociate entirely (it is not necessary for our purposes to explain why, but it is to do with the relative strengths of the interactions of water molecules with the acid molecules and with the H^+ and RCO_2^- ions formed). So, for example, when acetic acid dissolves in water some is present as undissociated CH_3CO_2H molecules and some dissociates to form H^+ and $CH_3CO_2^-$ ions.

$$CH_3CO_2H = H^+ + CH_3CO_2^-$$

The *total* amount of acid present, that is, the sum of CH_3CO_2H and H^+, is equal to the amount of acetic acid put into solution.

Now, the common measurement of acidity, pH, measures *only* the concentration of H^+ present in solution. The relationship between pH and H^+ concentration ($[H^+]$) is logarithmic and expressed by the equation:

$$pH = -\log[H^+]$$

(You will not be expected to use this!)

The pH scale generally extends from about 0 up to about 14: pH 7 is neutral; solutions with pH less than 7 are acidic (the lower the number the greater the acidity); solutions with pH greater than 7 are alkaline (the higher the number the greater the alkalinity and the weaker the acidity).

We can alter the pH of a solution of acetic acid by adding $CH_3CO_2^-$ ions in the form of potassium acetate. When $CH_3CO_2^-$ ions are added some will combine with H^+ ions to form CH_3CO_2H.

- ❍ What will happen to the pH of the solution when potassium acetate is added?
- ● The concentration of H^+ will go down, so the pH will increase.

Although the H^+ ion concentration goes down, the concentration of CH_3CO_2H will increase by an equivalent amount (because for every one H^+ ion removed one CH_3CO_2H molecule is formed). So, the *total* acid present – CH_3CO_2H plus H^+ – remains constant even though the pH changes.

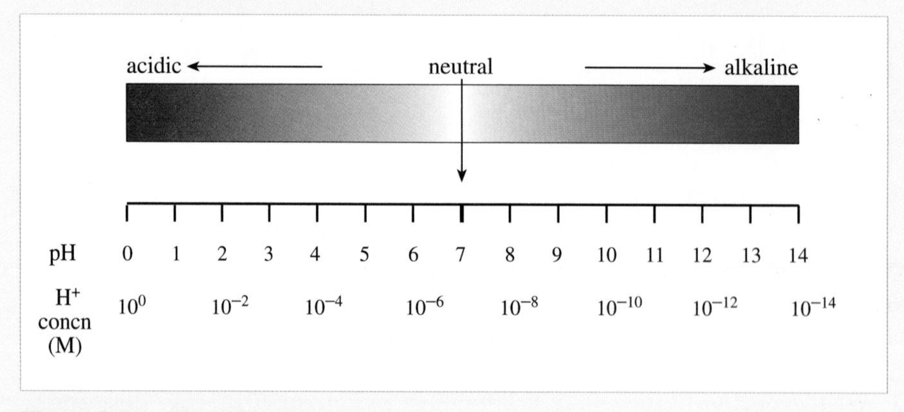

Figure 3.26 The pH scale.

Indeed, when acetic acid and hydrochloric acid are administered to the tongue at equal *total* acid concentrations then the chorda tympani responses are of roughly similar magnitude (Figure 3.24b). It would seem therefore that sour taste cells do not respond as pH detectors of the external environment, and that extracellular pH is not the sour stimulus. What then is?

As long ago as 1930 it had been suggested that:

> ... all acid solutions which taste equally sour produce the same pH *within the interior of the cell.* (our emphasis)
>
> N. W. Taylor, F. R. Farthing and R. Berman (1930) *Protoplasma*, **10**, pp. 84–97.

Remarkably, it would seem that this hypothesis is now proving to be correct. Take a look at Figure 3.27. This shows the interior pH changes of a taste receptor cell within a rat fungiform papilla when it is exposed on its extracellular side to acid stimuli.

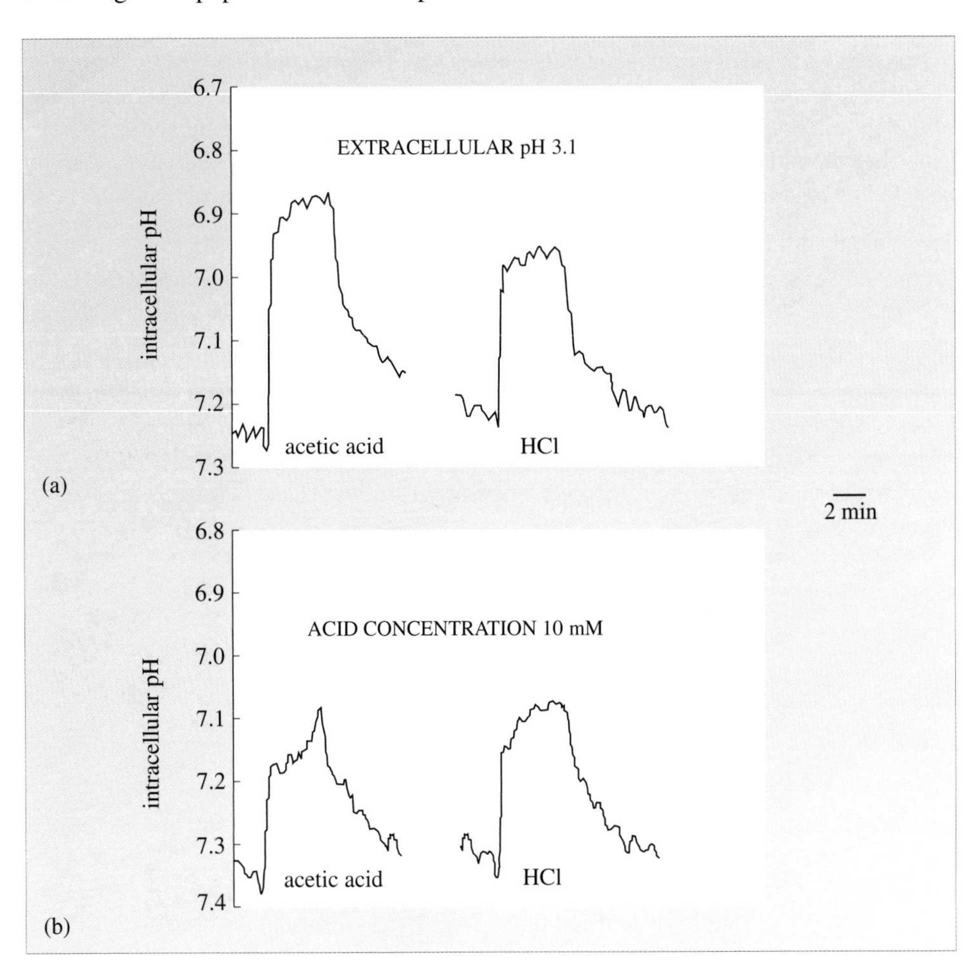

Figure 3.27 Intracellular pH changes of taste receptor cells when exposed to acid stimuli (a) of identical pH and (b) of equal concentration.

When acetic and hydrochloric acids of identical pH values are applied (Figure 3.27a), the intracellular pH of the taste cell decreases more in response to acetic acid than to hydrochloric acid. Additionally, when the two acids are applied at equal concentration (Figure 3.27b) then the changes in intracellular pH are of similar magnitude.

❍ What deductions can you make from the traces in Figure 3.24 and those in Figure 3.27?

- The general shapes and relative responses for the two conditions – acid stimuli applied either (a) at equal pH, or (b) at equal concentration – suggest that chorda tympani activity and intracellular pH are directly correlated.

Although one must be careful about linking two such observations as cause and effect, two additional pieces of evidence strengthen the connection. First, we noted earlier that carbonated water has a slightly sour taste. Does carbonated water reduce intracellular pH? Indeed it does: when exposed to a pH 7.4 bicarbonate/carbon dioxide solution (HCO_3^-/CO_2), the intracellular pH of sour taste cells falls from around 7.32 to 7.08. This is because inside the cell the enzyme carbonic anhydrase catalyses the conversion of carbon dioxide to carbonic acid (H_2CO_3) which in turn produces H^+ ions:

$$CO_2 + H_2O = H_2CO_3$$

$$H_2CO_3 = H^+ + HCO_3^-$$

When sour taste cells are subsequently exposed to a compound that blocks the action of carbonic anhydrase, the fall in pH is significantly reduced, to about 7.27. Observation of the chorda tympani activity showed that its activity was reduced by about 50 per cent when the carbonic anhydrase blocker was present. Since carbonic anhydrase activity is found *inside* the taste receptor cell, it does seem that intracellular pH is important.

The second additional piece of evidence that points to intracellular pH as the 'trigger' for the sour taste cell is the time dependence of chorda tympani activity and intracellular pH. These are shown for CO_2 in Figure 3.28; the two are reasonably well-correlated.

While 'sour' taste is brought about by a change in intracellular pH, it is currently not understood at the molecular level what the subsequent steps in the transduction process are. Presumably cell depolarization is brought about by the change in pH triggering intracellular secondary messengers, for example Ca^{2+} or cAMP, with depolarization effecting the release of a neurotransmitter. These pathways are for future research to elucidate.

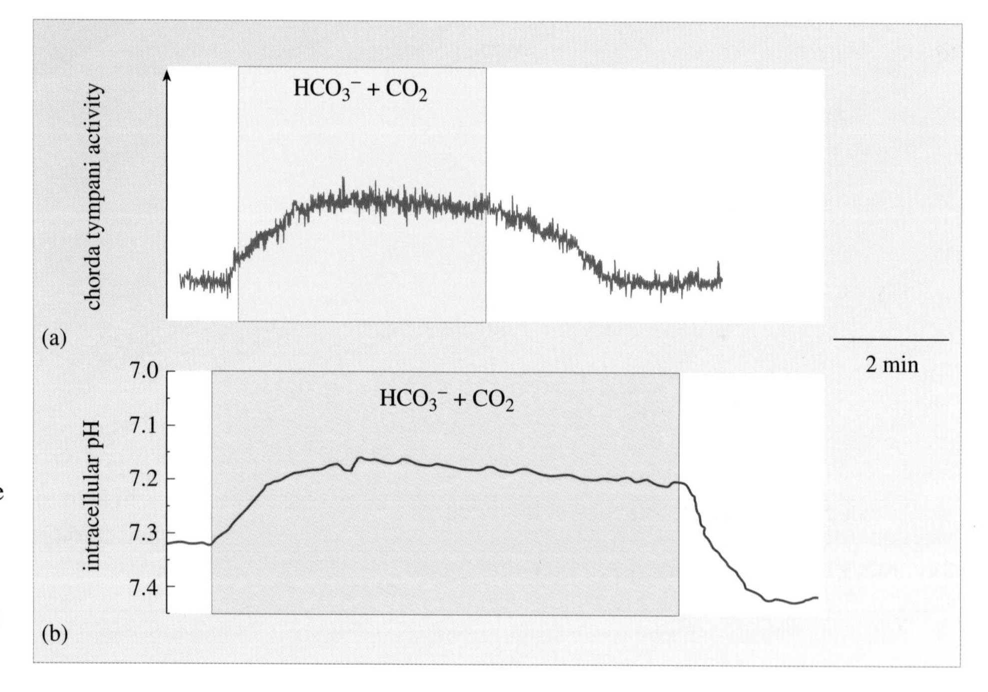

Figure 3.28 The time dependence of (a) chorda tympani activity and (b) intracellular pH of a sour taste (the extracellular pH is 7.4). The shaded regions represent the period of exposure to the CO_2 solution.

3.9 Salt taste

Our most common experience of salt taste is from sodium chloride, NaCl, the 'common' salt we use for flavouring food. Sodium chloride is an ionic substance; that is, the particles that constitute it are charged. Here the ions are the positively charged sodium ion, Na^+, and the negatively charged chloride ion, Cl^-. The saltiness perception we experience when tasting sodium chloride seems to be brought about by two possible pathways. Both involve transport of Na^+ across the cell membrane via a Na^+ ion channel to bring about cell depolarization. One involves Na^+ ion channels located in the part of the membrane of the taste cell that is exposed directly to mixtures of tastants. The second involves another type of Na^+ ion channel that is located in the membrane that butts up against other cells in the taste bud. Where two cells butt up together they form a **tight junction** between them, and tight junctions are permeable only to small ions and molecules. Large ions and molecules cannot gain access. This is important because the effects of two different types of Na^+ ion channel can be differentiated by the diuretic amiloride (Section 3.6), which is able to block accessible Na^+ ion channels.

❍ Which salt-taste Na^+ channels will be blocked by amiloride, those on the surface of the taste bud or those in the tight junctions?

● Amiloride will only be able to access those ion channels that are on the surface exposed to the pore of the taste bud.

What happens, then, when sodium chloride is tasted in the absence, then in the presence, of amiloride? Human subjects tasted four different solutions of sodium chloride. Each of these solutions were perceived to have additional taste qualities other than salty – notably sour, sweet and bitter – so the participants were asked to assess the magnitude of the taste intensities of all the perceived taste qualities. The data obtained are shown in Figure 3.29.

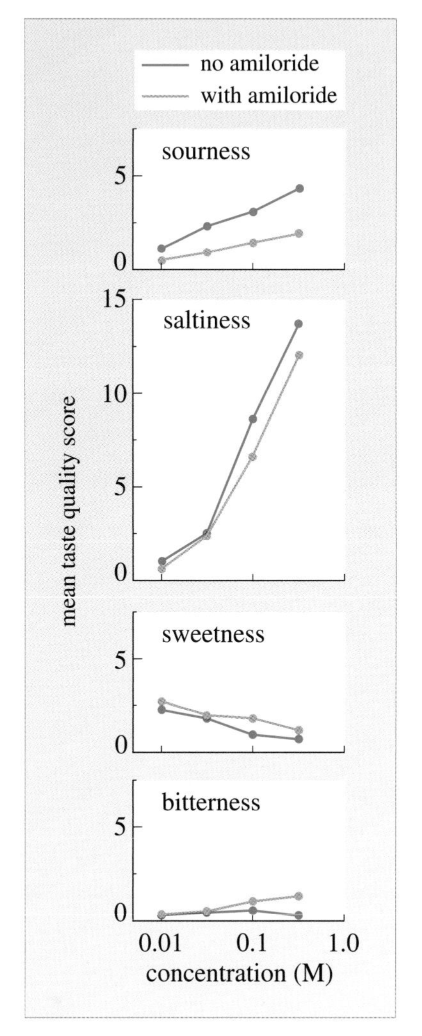

Figure 3.29 Perceived taste quality intensities of sodium chloride solutions in absence and presence of amiloride.

❍ Apart from a salty taste, what is the other major taste quality that sodium chloride has?

● At increasing concentrations the solutions have a perceptible sourness.

❍ What is the most intense taste perception of the weakest solution (0.01 M or 0.6 g l^{-1})

● The strongest perception is sweetness!

This is a well-recognized perception of dilute salt solutions, though the mechanism is not yet understood.

Participants were then asked to taste sodium chloride solutions to which 10 μM amiloride (2 mg l^{-1}) had been added (Figure 3.29).

❍ What effect does the presence of amiloride have on the saltiness of sodium chloride?

● It reduces the perception, but not by much.

❍ Which perception is most reduced by amiloride?

● The sourness of sodium chloride is most reduced.

These results seem to imply that the salt perception of sodium chloride is transduced more by the tight-junction ion channels, whereas the pore-accessible channels that are blocked by amiloride are responsible for some of the saltiness and most of the sourness.

Given that sodium chloride tastes salty, it seems a reasonable question to ask whether or not all salts taste salty, or whether other salts evoke the taste quality perceptions that for sodium chloride are minor components of its taste. Table 3.3 contains the perceived intensities of the component tastes for a variety of simple salts.

Table 3.3 Perceived taste intensity data for a range of different 0.1 M salt solutions.

Salt	Formula	Taste intensity			
		salty	sour	bitter	sweet
sodium chloride	$NaCl$	23	4	2.5	0.5
lithium chloride	$LiCl$	20	8	5	0
potassium chloride	KCl	9	3	25	0
calcium chloride	$CaCl_2$	6	0	46	1
ammonium chloride	NH_4Cl	20	5	14	0
sodium nitrate	$NaNO_3$	5	3	7	1
sodium acetate	NaO_2CCH_3	5	5	3	2

Focus your attention on the chloride salts first. Here the positively charged cation varies, from Na^+ through Li^+, K^+, Ca^{2+} to NH_4^+.

- ❍ Does the cation matter to salt taste perception?
- ● Yes it does. LiCl has an overall perception most similar to NaCl, but KCl and $CaCl_2$ are perceived largely as bitter and NH_4Cl is almost equally salty and bitter.

Now focus your attention on the sodium salts, where the Na^+ cation is constant but the negatively charged anion varies from Cl^- to NO_3^- and $CH_3CO_2^-$.

- ❍ Does the anion affect salt taste perception?
- ● Indeed it does. Whereas NaCl has a dominant salty taste, neither $NaNO_3$ nor NaO_2CCH_3 have dominant tastes. $NaNO_3$ is perceived as weakly salty and bitter, NaO_2CCH_3 is perceived as weakly salty and sour.

- ❍ Given what you know about how the salt taste may be transduced, and that K^+ and Ca^{2+} are large compared to Na^+ and Li^+, and NO_3^- and $CH_3CO_2^-$ are large compared to Cl^-, how might you account for the different taste perceptions of the salts in Table 3.3?
- ● The amiloride blocking experiment suggested that saltiness is more related to the ion channels in the tight junctions. Perhaps the larger ions are unable to access these, so the other taste perceptions become dominant over saltiness.

Although we still don't understand the mechanism of salt taste perception, both cation and anion are crucial to it. This is not altogether too surprising if, as is generally suspected, the salt taste quality has developed to respond to the body's electrolyte requirements.

3.10 Summary of Sections 3.8–3.9

Sour and salt taste qualities are transduced by ions traversing the taste cell membrane via ion channels. Sour taste has probably evolved to detect unripe and/or decaying foods. The sour taste is elicited by both organic and mineral acids, and taste cells appear to respond to changes in the intracellular pH. Salt taste probably exists to respond to changes in the body's electrolyte balance. It is elicited by ionic substances such as sodium chloride and potassium chloride. Salts other than sodium choride elicit additional taste qualities like sourness and bitterness.

3.11 Taste mixtures

So much for individual taste qualities. However, it is seldom that we are exposed to single tastants and single taste qualities. More commonly our 'world of taste' encompasses mixtures of tastants from all the taste categories. As you will probably have noticed from your everyday experience, when taste compounds are combined together they often interact in such a way that each tastes somewhat different than it does when tasted on its own. To begin to understand how this happens, mixtures containing two, sometimes three, components have been studied, most commonly using component concentrations that are above the threshold detection level. Mixtures of sweet tastants have been studied the most, umami the least. Since different tastants of any one taste quality, for example bitter, can each interact differently with tastants of other taste qualities, it becomes apparent the research required to understand this one area of human taste is vast.

Several kinds of taste interaction have been observed: enhancement, synergy, suppression and masking.

Enhancement

This is the kind of interaction observed when two or more tastants are mixed together at above threshold concentrations, resulting in a particular taste quality being increased in intensity.

Enhancement is relatively rare between tastants of different qualities but tends to be a general observation for tastants of the same quality. An example of enhancement across taste qualities is observed between sodium choride (salty) and the amino acid arginine (which has both bitter and umami qualities); arginine, though it has no salt taste quality of its own, enhances the salty taste of sodium chloride.

Synergy

Synergy reflects a stronger form of positive interaction, but it is rarely observed in taste perception. One example involves the combination of MSG and 5′-ribonucleotides, which synergize their respective umami tastes; a second involves the sweeteners aspartame and acesulfame, which synergize sweetness qualities. So far, synergy across taste qualities has not been observed.

Suppression

Suppression is the reverse of enhancement; that is, when tastants are mixed together a particular taste quality is decreased. It is a common phenomenon, especially among tastants of different taste qualities. Lemon squash benefits from the sweetness of sucrose suppressing the sourness of the citric acid in the lemons, as well as from the sourness of the lemons suppressing the sweetness of the sugar. This kind of symmetrical suppression is common, but asymmetrical suppression is also known. For example, in mixtures of quinine and sodium chloride the saltiness is unaffected by quinine whereas the bitterness of the quinine is suppressed by 50–70 per cent depending on the sodium chloride concentration used.

Masking

Masking, the reverse of synergy, is a stronger form of negative interaction. One example involves mixtures of sodium chloride and urea in which sodium chloride masks the bitterness of urea. Another involves gymnemic acid, a component of the plant *Gymnema sylvestre,* which masks the sweetness of any sweet tastant. Gymnema tea is an Ayurvedic remedy for diabetes.

Interesting effects can be observed when a third tastant is added to a two-component mixture. When the bitter-tasting urea is mixed with the sweetener sucrose, mutual suppression – the sweet suppressing the bitter and vice versa – occurs. When salt is added to the bitter–sweet mixture, it has a large masking effect upon the bitter taste but a weak suppressive effect upon the sweet. When released from the suppression by the bitter, the sweetness will increase in intensity and the perception of the three-component mixture is predominantly sweet.

One of the more curious observations of taste interactions relates to the perceived detection thresholds of mixtures of tastants. The American scientist Joseph Stevens has investigated how mixtures of tastants can have perceptible tastes even when *each* of the components is present at a concentration that is too low for it to be detected separately. Stevens' method involves comparison of the detection thresholds of each component determined separately with the detection thresholds of mixtures of the components. In these mixtures the components are present in the same ratios as their individual detection thresholds, but each mixture is of a different overall concentration. A representative set-up for studying a two-component mixture is shown in Table 3.4.

Table 3.4 Hypothetical series of solutions used to determine the detection threshold of a two-component mixture, comprising component A (individual threshold 20 mM) and component B (individual threshold 8 mM).

Solution	Concentration of component A / mM	Concentration of component B / mM	Ratio of A / B
1	20	8	2.5
2	15	6	2.5
3	10	4	2.5
4	5	2	2.5
5	2.5	1	2.5

Stevens used mixtures of three, six, twelve and twenty-four tastants. Table 3.5 shows data obtained for the sweet tastant aspartame, though analogous data were obtained for all tastants studied including examples of bitter, sour, salty and umami.

Table 3.5 Concentrations of aspartame in solutions of mixtures that had just-perceptible tastes.

Number of components	Aspartame concentration in mixture / μM
1†	17.6
3	7.4
6	3.7
12	1.65
24	1.20

† aspartame alone

❍ What do you notice about the concentration of aspartame present in the mixtures as the number of components increases?

● As the number of components increases, the aspartame concentration decreases.

This implies that there must be some kind of interaction between the separate taste modalities. You will be able to appreciate the kind of interaction by carrying out the following calculation.

❍ Use the aspartame concentrations to calculate the factor for each mixture by which the aspartame concentration decreases.

● For the three-component mixture the factor is 17.6/7.4, which is 2.4. The other factors are: 17.6/3.7 = 4.8 for the six-component mixture; 17.6/1.65 = 10.7 for the 12-component mixture; and 17.6/1.20 = 14.7 for the 24-component mixture.

❍ What do you notice about these dilution factors and the number of components in each mixture?

● The dilution factor is proportional to the number of components, and apart from the twenty-four-component mixture, to a reasonable approximation the dilution factor and the number of components are the same.

Consequently, taste is a highly integrative, or additive, sense across each of the taste qualities. How this is achieved is currently unknown, but these observations suggest some kind of neural integrator across quality channels.

Another observation related to threshold detection levels relates to the ability to detect a particular taste quality in the presence of an above-threshold concentration of a different taste quality. For example, the detection threshold of sodium chloride rises by a factor of three to four if either sucrose or citric acid is present as a 'suppressing' taste.

❍ How might you expect the detection threshold of sodium chloride to change if *both* sucrose and citric acid are present?

● If they work in concert, the change might be nine to sixteen-fold.

Indeed, the effect is about a factor of ten. Of course, this is what happens in our everyday lives; we often need to detect a taste in the presence of some background taste or flavour, such as when we season a complex food by adding salt. What this observation seems to imply is that a taste is far more efficiently masked by the addition of multiple tastants rather than just one.

Is it possible, then, to answer questions such as 'How do sour and bitter tastants interact in mixtures?' or 'Does saltiness increase sourness?' A quick look at Table 3.6 would suggest that, for the moment, it is not. Whether or not suppression or enhancement is observed appears to depend on the particular tastants examined and the concentrations at which they are used.

Table 3.6 Taste perception of mixtures of tastants.

Components (threshold intensity level)		Outcome	Comment
sodium chloride (below)	citric acid (below)	slightly synergistic	
sodium chloride (below)	citric acid (above)	suppression	sourness of citric acid reduced
sodium chloride (above)	citric acid (above)	enhancement	citric acid enhances saltiness of sodium chloride; sodium chloride enhances sourness of citric acid
sodium chloride (well above)	citric acid (well above)	suppression	citric acid reduces saltiness or has no effect; sodium chloride reduces sourness
sodium chloride (well above)	HCl (well above)	no effect	sourness of HCl unaffected; saltiness of sodium chloride unaffected
caffeine (below)	citric acid (above)	suppression	caffeine suppresses sourness
caffeine (above)	citric acid (below)	suppression	citric acid suppresses bitterness
caffeine (well above)	citric acid (well above)	enhancement	citric acid strongly enhances bitterness

3.12 The coding of taste

The taste cells that detect tastants in our mouths are but the first stage in the neural pathway that links a gustatory stimulus to its taste perception (Figure 3.30).

As described in Section 3.1, taste cells, unlike olfactory receptor cells, are not neurons; they are modified epithelial cells that synapse with the first neurons of the gustatory system. In the anterior two-thirds of the tongue these are the peripheral neurons that form the chorda tympani, in the posterior one-third taste cells are innervated by neurons of the glossopharyngeal nerve.

These cranial nerves channel signals to the nucleus of the solitary tract where they are fed to areas of the brain associated with the processing of gustatory information.

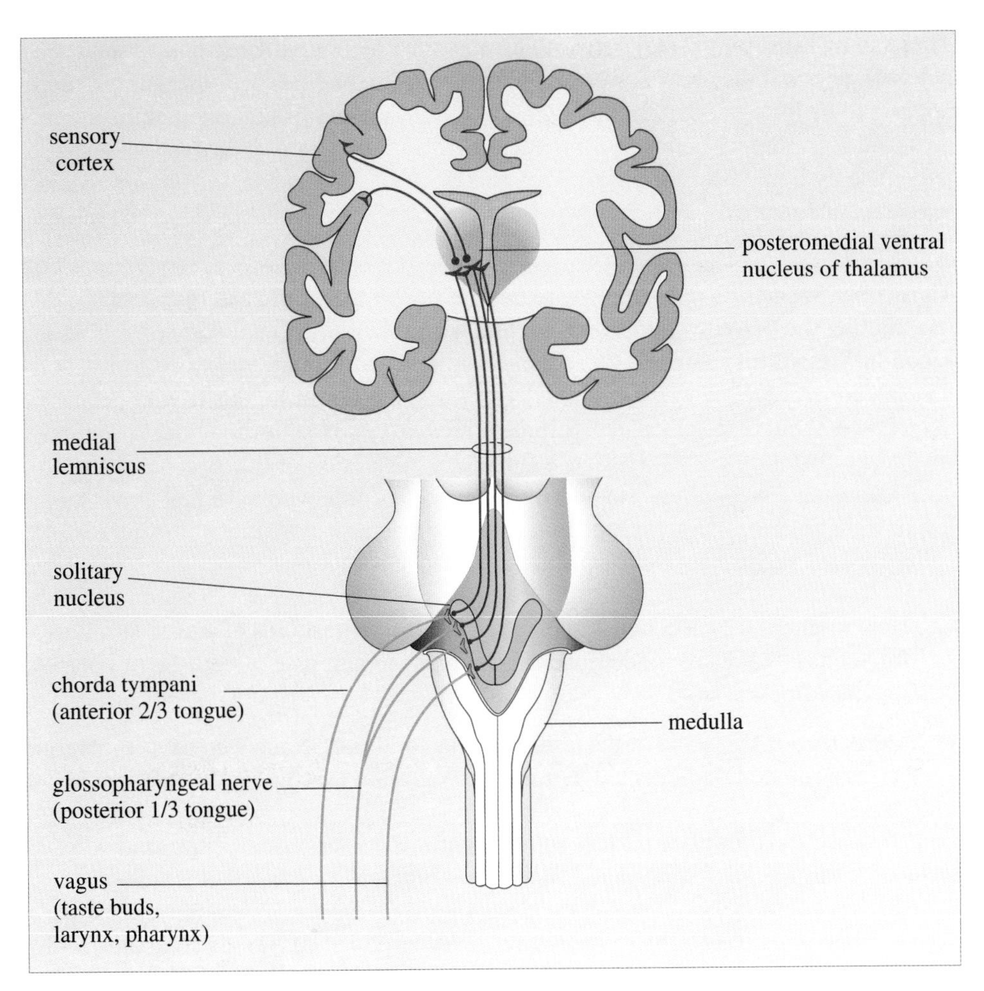

Figure 3.30 The gustatory neural pathway.

Just precisely where these taste-responsive regions of the human brain are, and to which type of taste they respond, has begun to be uncovered by functional imaging using fMRI and PET techniques. Figure 3.31 shows a composite fMRI scan for seven subjects showing their responses to the tastes elicited by 1 M glucose and 0.1 M sodium chloride.

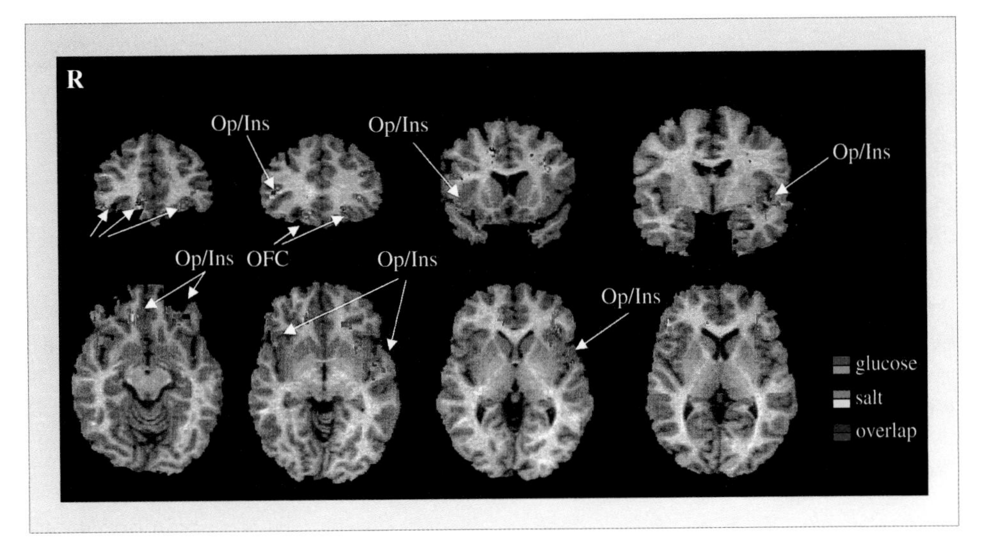

Figure 3.31 Composite fMRI images of brain areas activated by glucose (green/light-green areas) and sodium chloride (orange/yellow) areas. Areas of overlap are coloured blue. (OFC, orbitofrontal cortex; Op/Ins, frontal operculum and insula).

These show the orbitofrontal cortex to be activated by both glucose and sodium chloride, and that the areas activated by the two taste qualities are adjacent but there appears to be little overlap. In the region of the insula and frontal operculum, some areas are activated by both stimuli while others only by glucose. Activation of the amygdala by both taste stimuli can also be observed: for glucose, the left amygdala was activated in most subjects; for sodium chloride, activation was observed in just over half the subjects, but no clear regional pattern could be detected.

These observations have been interpreted in terms of the insula and operculum constituting the **primary gustatory cortex**, with the orbitofrontal cortex forming the **secondary gustatory cortex**. As its name implies, the primary gustatory cortex is the first cortical area where a stimulus is represented. For taste, this is where *detection* is thought to take place. Stimulus *recognition* is believed to occur in the secondary gustatory cortex. Evidence for such a separation of function has been obtained using psychophysical and PET studies of people who have had either the left or right anterior temporal lobe removed during surgical treatment for intractable epilepsy. This surgery leaves the primary gustatory cortex intact.

❍ How might you expect detection and recognition thresholds of a taste stimulus for subjects who have undergone such surgery to compare with those of control groups who have not?

● The detection threshold of the taste stimulus for those people who have undergone surgery should be unaffected, but their recognition thresholds should increase.

Indeed, when their ability to detect and recognize the sour taste of citric acid was tested, the three groups were found to have similar detection thresholds (about 9×10^{-5} M) but the group that had the right anterior temporal lobe removed had a significantly impaired ability to recognize the taste (2.8×10^{-3} M, as compared with 5×10^{-4} M for either the control group or those who had their left anterior temporal lobe removed). When the control group were analysed for brain activity using PET, the area of the brain that was activated most strongly was the right medial orbitofrontal cortex, though the left caudolateral orbitofrontal cortex and right anteromedial temporal lobe were also found to be centres of activity (Figure 3.32). Thus, surgical loss of a significant part of the right anterior temporal lobe would be consistent with impaired taste recognition if this function is the role of the secondary gustatory cortex.

So much for the levels at which gustatory information is handled. But how is that information coded?

Olfaction, where there are thousands, if not millions, of different odours and several hundred different types of odorant receptors, uses a combinatorial mechanism that involves generating a pattern of activity that is interpreted by the olfactory cortex. Gustation, in contrast, seems to involve a small range of taste qualities – currently totalling five – and a similarly small number of taste receptors and ion channel types. Does that mean that taste is coded differently from olfaction, perhaps by a labelled-line mechanism? This has been a topic of intense debate for some decades and is still not entirely resolved. In fact, how gustatory information is coded is unknown, although our current understanding points to a coding mechanism that involves a pattern of activity across nerve cells but in which certain types of cell are crucial to the perception of taste modalities.

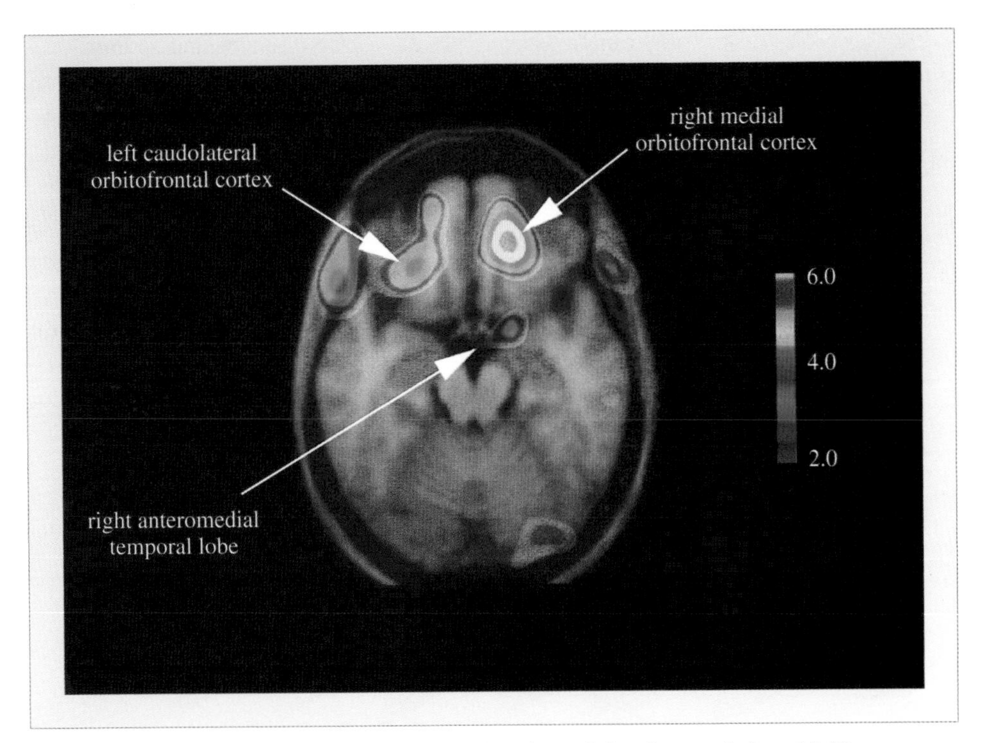

Figure 3.32 PET scan showing regions of brain activity due to citric acid. The centres of activity seen outside the brain are artefacts due to the mouth movement needed to perform the task.

Given the distribution of taste papillae, and the different innervating neurons, in the oral cavity, one potential representation of taste could involve a **topographic** or **chemotopic map**. This 'map' of the oral cavity associates regions of the mouth with different taste qualities. Indeed, the somatosensory response of the tongue and oral cavity is topographically mapped to Brodmann's area 3b of the cortex, the tongue in particular being represented in the postcentral gyrus. However, there is little evidence for the reality of a taste 'map', despite such a map appearing in many books. In fact, although there are some regional differences in the patterns of sensitivity to tastants, in humans these are relatively small and most taste modalities elicit rather similar responses from all areas that house taste cells (Figure 3.33 overleaf).

A second possibility is that the taste cells themselves respond only to particular taste qualities and therefore code for them directly. There are only limited data regarding the sensitivity of individual taste cells, though it is believed that a single taste cell can respond to more than one type of taste quality. However, this may not be the case for some taste cells that are responsive to sweet and bitter. For example, the T1R sweet taste receptors are expressed in 15–30 per cent of circumvallate, foliate and palate taste buds. This appears to be a similar distribution to that of the bitter taste T2R receptors so it seems reasonable to ask whether or not there is any overlap between the two taste cell populations. When different fluorescent probes are used to label the two types of receptor, pictures such as that shown in Figure 3.34 (overleaf) are obtained.

❍ What do you notice about the two populations of cells in Figure 3.34?

● Cells expressing bitter taste receptors are separate from those expressing sweet taste receptors.

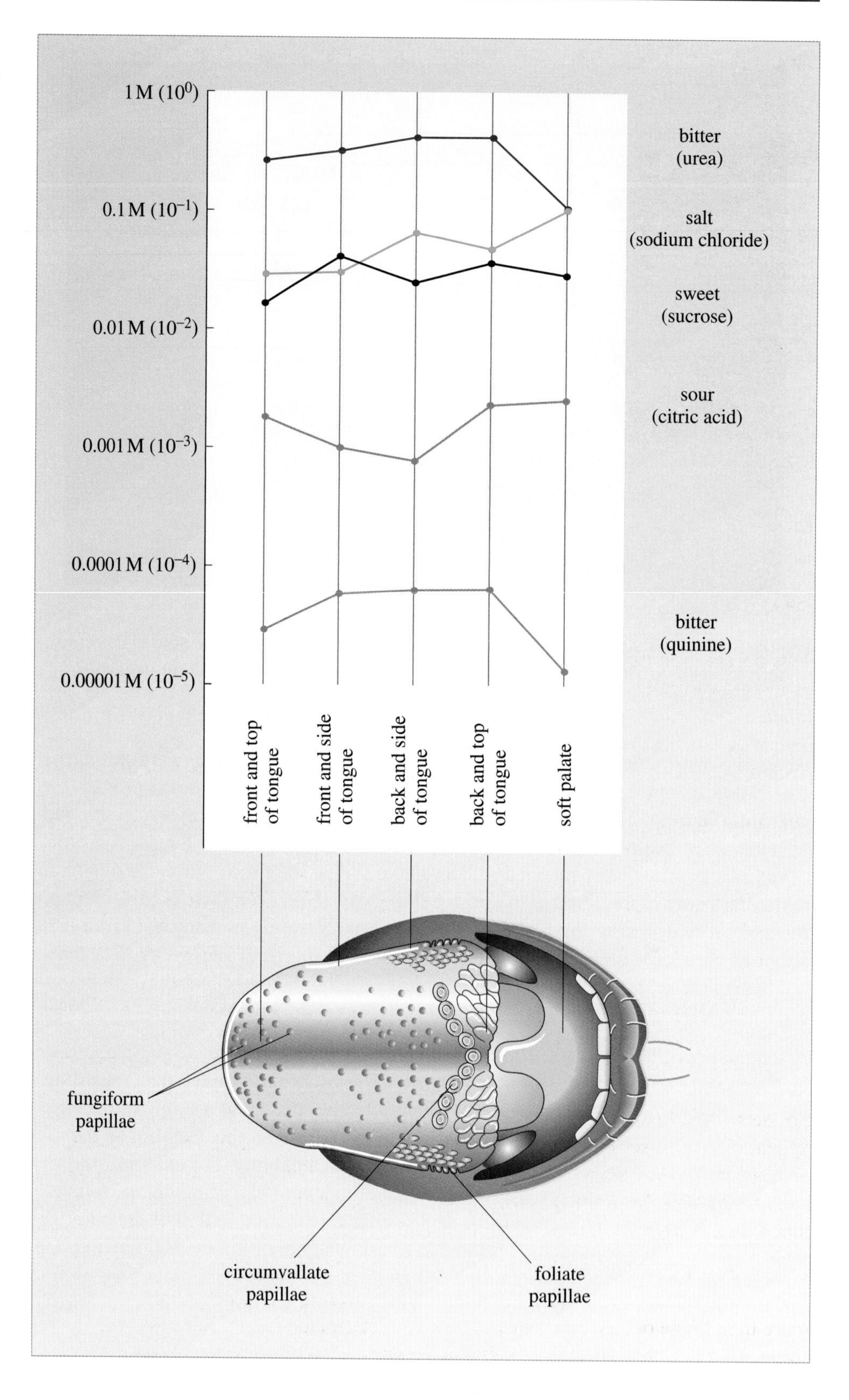

Figure 3.33 The responsiveness of regions of the mouth to tastants of four taste qualities.

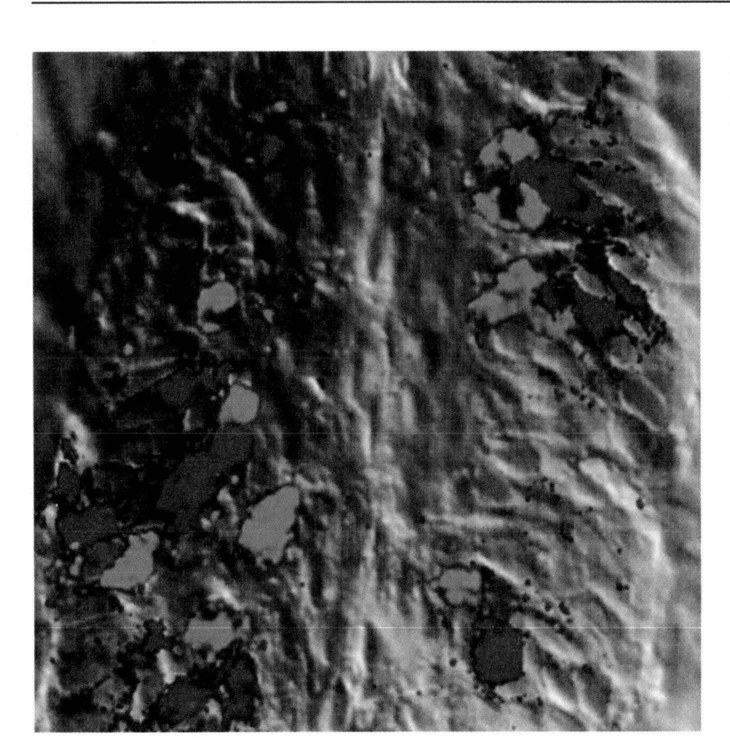

Figure 3.34 Fluorescence labelling of the cells in a circumvallate papilla to show cells expressing sweet taste receptors (green) and bitter taste receptors (red).

This observation implies that sweet and bitter tastes are encoded by the activation of different taste cell types. It is not known if this is a general phenomenon for all taste modalities, but, even if it is, it does not extend to the innervating and subsequent neurons of the gustatory pathway. Figure 3.35 (overleaf) shows the activity of fibres from the chorda tympani and glossopharyngeal nerve to the familiar taste qualities.

❍ What similarities and differences do you notice about the two patterns of activity in fibres of the chorda tympani and glossopharyngeal nerve?

● First, all nerve fibres exhibit activity to each tastant. Second, generally each fibre shows greater activity to one type of tastant. Third, in either nerve certain taste qualities have higher overall activities than others; for example, sour and bitter tastants elicit much stronger responses (≥ 40 spikes s^{-1}) than sweet and salt (≈ 10 spikes s^{-1}) in the glossopharyngeal fibres. Fourth, fibres from the different nerves have different sensitivities and profiles to the various taste qualities, the chorda tympani, for example, being more responsive to sweet and salt tastants.

So, these peripheral fibres are not uniquely tuned to one taste quality, though they do appear to exhibit selective responses. However, the peripheral fibres converge onto neurons in the brainstem. Recordings of impulses from neurons in the solitary tract and also of the orbitofrontal cortex have revealed that they, similarly, are broadly tuned; that is, they respond to tastants of all taste qualities (Figure 3.35). This broad tuning appears to be characteristic of neurons at every level of the gustatory system.

Although broadly tuned, these neurons generally respond to one taste quality more than to the others, and they are therefore often labelled as 'salt best' and so on. That does not mean that they alone carry the salt taste code. For example, Figure 3.36 (overleaf) shows the response patterns of sixteen hamster 'salt best' solitary tract neurons to 0.01 M sodium chloride – the stimulus these respond to – and to 0.1 M and 1 M potassium chloride solutions.

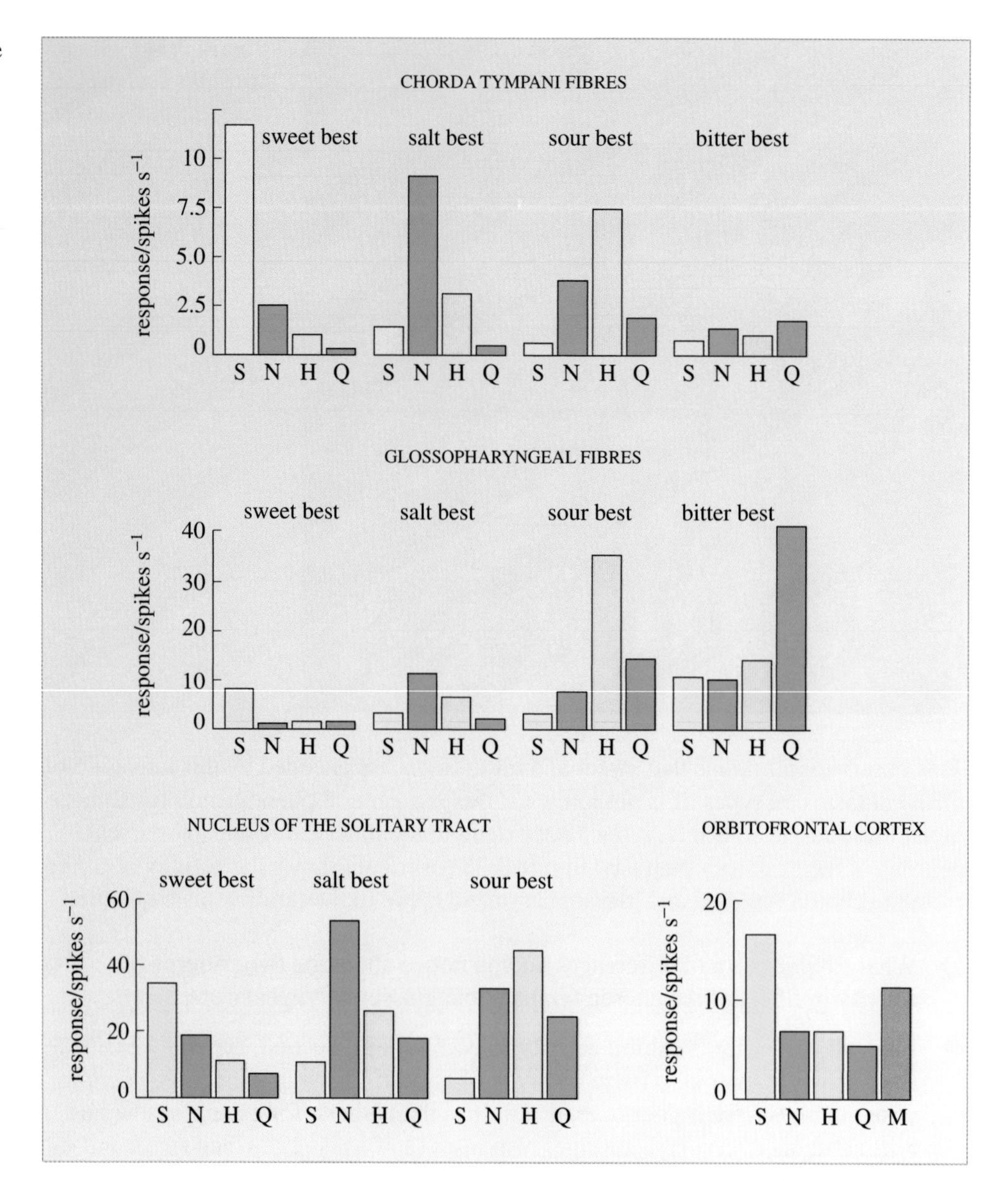

Figure 3.35 The tastant response profiles of different types of fibres of the chorda tympani and glossopharyngeal nerve, of three different types of neuron of the solitary tract, and of a neuron of the orbitofrontal cortex: S, sucrose; N, sodium chloride; H, hydrochloric acid; Q, quinine; M, monosodium glutamate.

❍ What do you notice about the patterns for KCl as compared with that for NaCl?

● While the pattern for 0.1 M KCl is rather different to that for 0.01 M NaCl, that for the 1 M solution is remarkably similar.

The implication is that at the higher concentration of KCl this population of 'salt best' cells would have difficulty discriminating between the two tastants. Since the two tastants *are* perceived as different, other neuron types must be contributing to the taste code.

Additionally, it is clear from the responses of the solitary tract neurons in Figure 3.35 that HCl evokes substantial activity in 'salt best' neurons. Now, neuronal activity is dependent on the tastant concentration, so, in a 'salt best' neuron, it is possible to achieve the same level of response from a high concentration of HCl as it is from a low concentration of NaCl. Yet these two tastants are clearly discriminable as having sour and salt tastes, respectively. In fact, rats conditioned to avoid NaCl do not avoid

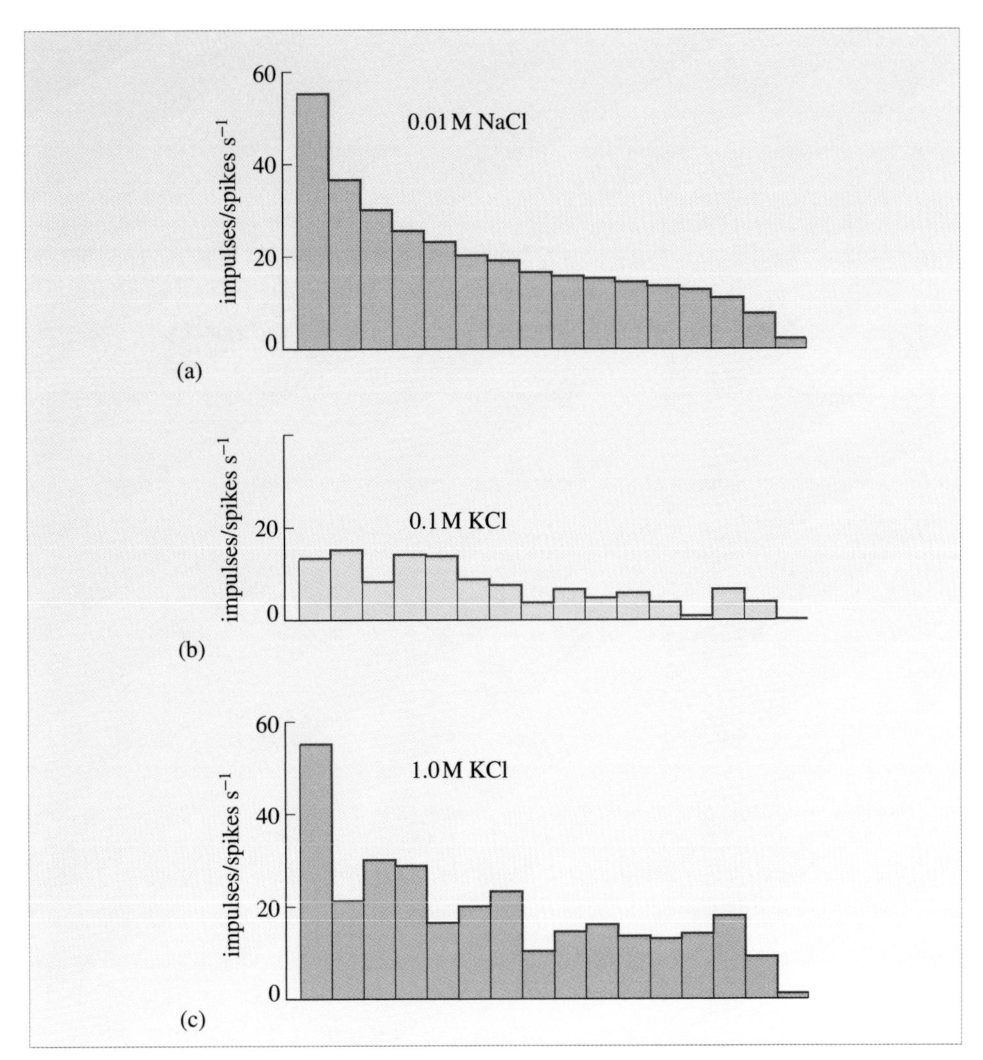

Figure 3.36 The responses of sixteen 'salt-best' neurons of the hamster solitary tract to 0.01 M NaCl, 0.1 M KCl and 1 M KCl.

HCl, nor do they avoid NaCl after conditioning to avoid HCl. We are therefore led to the likelihood that the taste is coded by a pattern of activity across cells. This is most clearly seen in Figure 3.37 (overleaf), which shows the patterns of activity of a range of solitary tract neurons – including 'salt best', 'sweet best' and 'sour best' – to 0.1 M NaCl and 0.1 M KCl solutions in the absence and presence of the sodium channel blocker amiloride.

When amiloride is absent the response profiles are very different, particularly for the 'salt best' neurons. These two tastants are easily discriminated under this condition. In the presence of amiloride, behaviourally the ability to discriminate between KCl and NaCl is lost. Looking at the response profiles when amiloride is present, these appear remarkably similar. Although it looks as if only the 'salt best' cells are affected, in fact the activity in all cells is reduced, even the 'sweet best' and 'sour best'. Of course, the 'salt best' neurons are most affected, and it is likely that these are the cells that define the predominant code for salt.

Taste, then, looks as if it is coded by the pattern of activity across the range of neurons, with some neurons playing a stronger role in defining the similarities and differences between patterns. However, our understanding of gustatory coding is

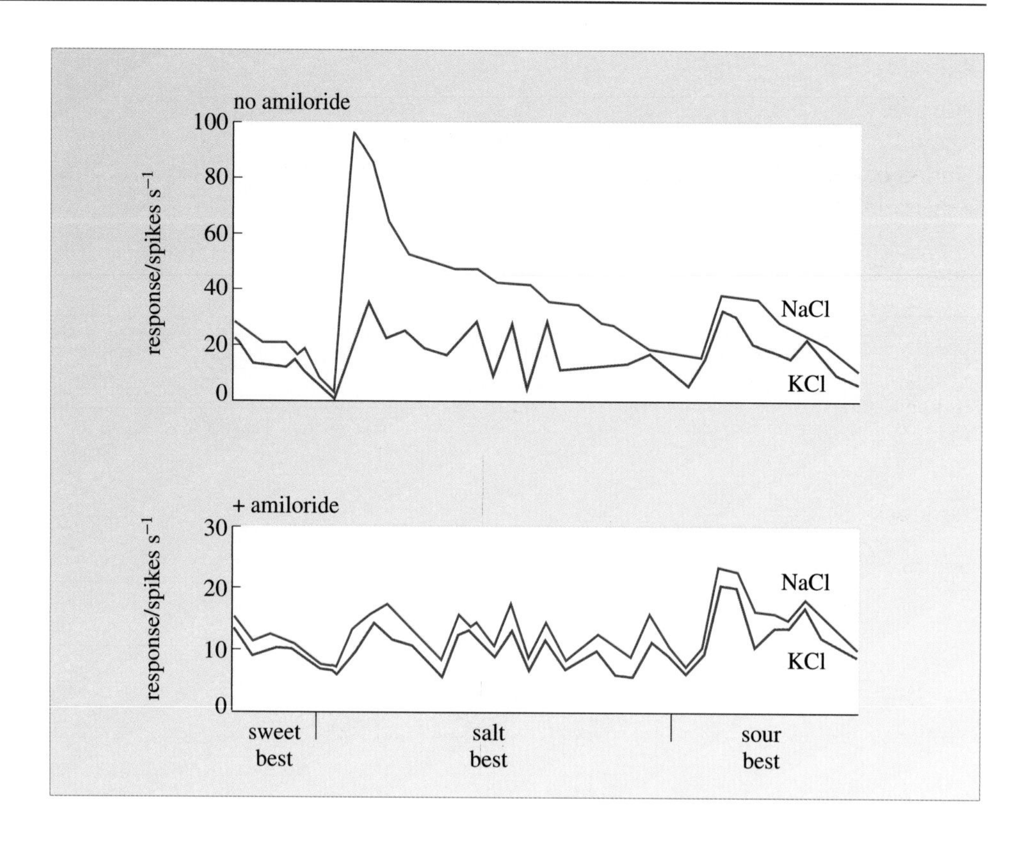

Figure 3.37 The response profiles of different solitary tract neurons to NaCl and KCl.

still in its infancy and it is likely that it will change rapidly in the coming years. We are still some way from knowing how the brain reads the code and represents it as a perception of a taste quality.

You should now read Chapter 26 of the Reader, *Taste perception* by Steve Van Toller.

3.13 Summary of Sections 3.11–3.12

Mixtures of different tastants can exhibit a range of taste interactions, including enhancement, synergy, suppression and masking. There is no obvious means of predicting the kinds of response that will be observed; this appears to vary depending on the tastant and its concentration. The threshold detection limit of a single tastant in a mixed solution decreases in inverse proportion to the number of components in the mixture.

The insula and the operculum form the primary gustatory cortex, the secondary gustatory cortex being the orbitofrontal cortex. Taste detection occurs in the primary cortex, taste discrimination in the orbitofrontal cortex. The peripheral nerve fibres and subsequent neurons of the gustatory pathway are broadly tuned and respond to tastants from a range of taste qualities. However, they generally respond to one taste quality more than the others and are therefore labelled 'salt best', etc. Taste appears to be encoded by the overall pattern of activity across the neurons rather than by activating specifically tuned sets of neurons.

Question 3.1

How might the presence of amiloride affect the perception of LiCl and KCl (you will need to make use of the data in Table 3.3)?

Question 3.2

Using the threshold detection levels contained in Table 3.7, what would be the required concentrations of quinine, sucrose, citric acid and saccharin to produce a solution of detectable taste containing the four components present in the same ratio as their individual detection thresholds?

Table 3.7 Threshold detection levels of four tastants.

Tastant	Threshold / μM
quinine	1.4
sucrose	640
citric acid	70
saccharin	10

Question 3.3

It is often implied that the back of the tongue is primarily the region for responding to bitter tastes. What evidence is there for this in Figure 3.33?

Question 3.4

It is often observed that those people who have had surgery in which the chorda tympani to each side of the tongue is severed, retain their perception of, and sensitivity to, all taste modalities. From what you understand of the gustatory system explain why this is so.

Question 3.5

Why does common salt have taste but no smell?

Question 3.6

The sweetness of sucrose is perceived as more intense in the presence of low concentrations of sodium chloride (10^{-2} M), but less intense at higher concentrations (10^{-1} M) of sodium chloride. Describe the taste interactions present in these two situations and suggest a possible reason for these observations.

Question 3.7

'Each sensory taste cell responds to one tastant only'. Comment on this statement.

Question 3.8

All taste qualities are receptor mediated. Is this true or false?

Question 3.9

The chorda tympani responds to sweet and salt tastes more than does the glossopharyngeal nerve. How accurate is this statement? Give your reasons.

Question 3.10

Is PROP taster status consistent with the idea that bitter is tasted at the back of the tongue?

4 Flavour: a merging of the senses

> … smell and taste are in fact but a single sense, whose laboratory is the mouth and whose chimney is the nose …

So wrote Brillat-Savarin. Today, we would call that 'single sense' flavour. When food is chewed and swallowed, odorants are released in the throat enabling them to reach the nasal cavity where they stimulate olfactory receptors, and tastants are released that stimulate the taste receptors on the tongue and in the mouth. What the brain perceives as 'flavour' is a combination of the sensory inputs from taste and smell, together with the tactile sensation of the food (such as the 'smoothness' of cream or the crunch of breakfast cereals), the sensation that comes from the stimulation of the trigeminal nerve (such as the 'hot' sensation of chilli, or the 'cool' sensation of mint) and temperature (think how differently you perceive a cup of hot coffee compared to a cup of cold coffee). Here, though, we shall discuss flavour in the more limited terms of the connectedness of taste and odour.

In our everyday lives we often use the terms taste and flavour interchangeably. However, we know from experience that the two *are* somewhat different. The perception of a cup of coffee is significantly different if we sip it with our nostrils open from when we sip it pinching our nostrils shut (admittedly not something we do often!). In the latter, the sensation is of a bland, somewhat bitter, taste due to the presence of the bitter tastant caffeine; in the former we experience a much more complex sensation. A similar limitation of perception is experienced when we have a cold and our nasal passages are blocked by mucus.

❍ To what would you ascribe the differences between the two perceptions?

● When the nostrils are blocked our ability to perceive the aromas of the volatile odorants present in the coffee drink is removed.

That flavour requires both taste and smell can be seen from Figure 4.1 (overleaf). This shows the difference in the ability to identify some common food items by taste alone and by taste in combination with smell. Whereas some flavours are more easily identified than others, all are better identified using the two senses combined. Indeed, coffee is remarkable in that it seems that it cannot be identified by taste alone.

Several human psychophysical experiments have demonstrated this interaction between smell and taste. One such experiment explored the intensity of the umami flavour brought about by MSG on its own and by MSG in the presence of 2-methyl-3-furyl disulfide (Figure 4.2a overleaf), a compound that has a garlicky/meaty odour.

Subjects were exposed in random sequence to solutions of four different concentrations of MSG both with and without 2-methyl-3-furyl disulfide. The umami flavour intensity of each solution was assessed on a scale of 0–100, 0 being very weak and 100 very intense. The intensity data in the absence and in the presence of 10 parts per million (p.p.m.) of the disulfide odorant are shown in Figure 4.2b.

❍ From the figure, what can you deduce about the relative intensity of the perception in the absence and presence of the disulfide?

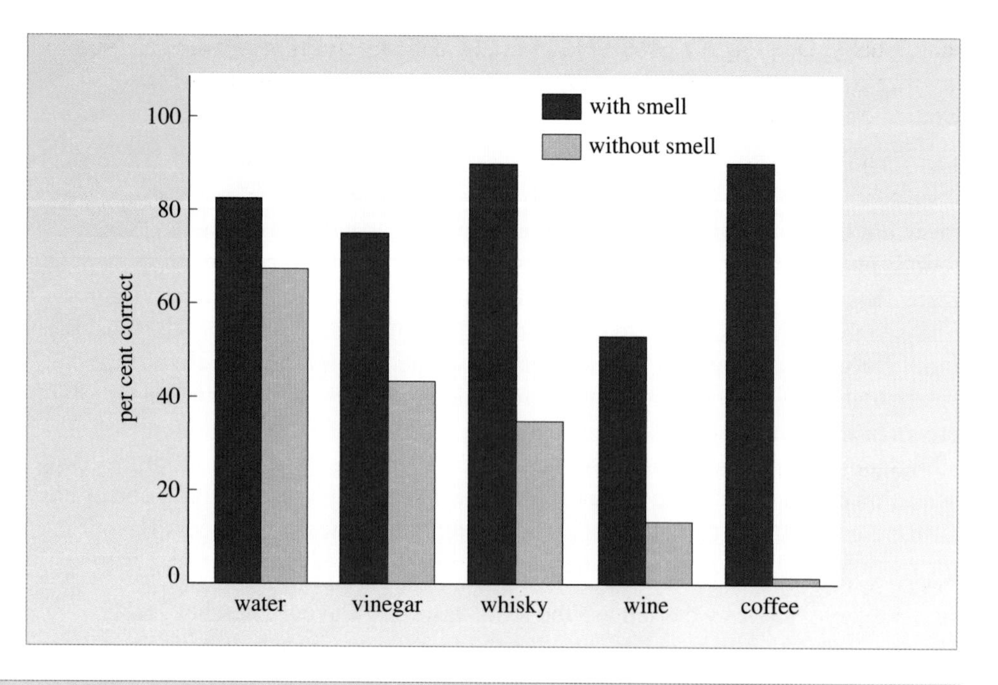

Figure 4.1 The use of smell and taste to identify some common foods.

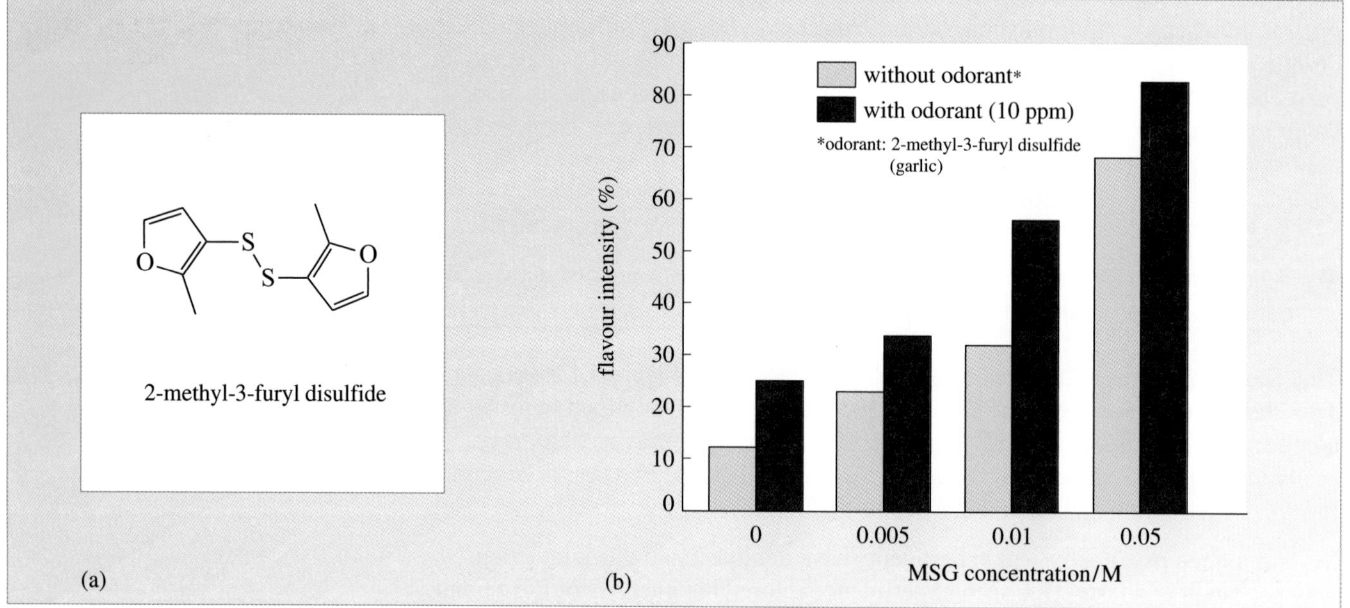

Figure 4.2 (a) The structure of 2-methyl-3-furyl disulfide, and (b) the intensity rating of savoury flavour to four solutions of MSG in the presence and absence of 2-methyl-3-furyl disulfide.

- The intensity is greater when the odorant is present. Furthermore, the intensity enhancement appears to be approximately constant across the MSG concentration range.

In other words, the flavour of umami appears to be *additive* between the taste perception – brought about by MSG – and the smell perception – elicited by the disulfide odorant. This additive interaction between taste and smell has also been investigated in an alternative way. As you have already seen, mixtures of different tastants interact in such a way as to lower the threshold detection level of the mixture components as compared with the threshold detection levels of the tastants individually. One might reasonably inquire if such an effect is observed for a mixture containing an odorant and a tastant. If it is, then crossmodal interactivity would be implied.

Just such an experiment has been performed using mixtures of the odorant benzaldehyde, which smells of almonds, with either the non-smelling, sweet tastant saccharin or the non-smelling, umami tastant MSG. Participants were subjected to four different conditions:

- benzaldehyde odour on its own;
- benzaldehyde odour while holding 10 cm^3 of water in the mouth;
- benzaldehyde odour while holding 10 cm^3 of solutions containing sub-threshold concentrations of saccharin in the mouth;
- benzaldehyde odour while holding 10 cm^3 of solutions containing sub-threshold concentrations of MSG in the mouth;

and the threshold detection levels of benzaldehyde were determined. The results are shown in Figure 4.3.

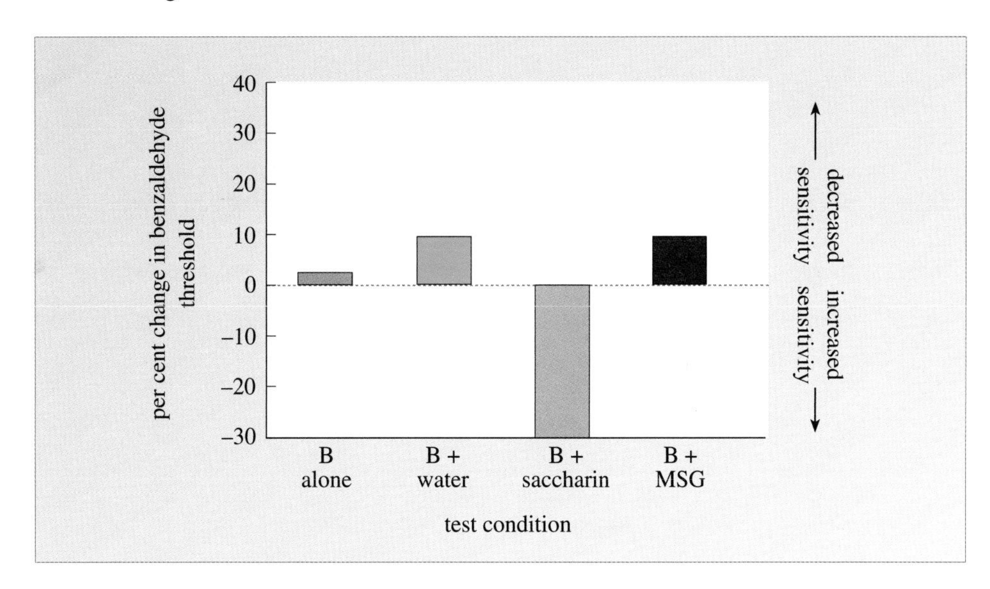

Figure 4.3 Change in benzaldehyde (B) threshold detection levels in the presence of water, saccharin and MSG.

❍ Compare the differences between the thresholds found when water and saccharin are present. What inference can you make? Why is water a better control condition than simply determining the threshold without anything in the mouth?

● The threshold detection level for benzaldehyde is lowered significantly in the presence of saccharin by approximately 40 per cent. Thus, it would appear that the sweet taste of saccharin is somehow adding to the sensory effect of benzaldehyde. The solution of saccharin can elicit a somatosensory, as well as a taste, sensation. Holding an equivalent volume of water in the mouth allows for the somatosensory element.

Because the mixture contains two components, if the crossmodal interaction behaves like that for a same-modal interaction (as occurs for taste) then the threshold detection level should be reduced by 50 per cent. The 40 per cent reduction observed for the benzaldehyde/saccharin mixtures implies that the interaction is indeed an additive one, though perhaps not entirely so.

❍ In Figure 4.3, what do you notice about the benzaldehyde threshold in the presence of MSG?

● There is essentially no difference in the benzaldehyde detection threshold between a solution of MSG and water. So MSG has no effect on benzaldehyde detection.

So, the additive interaction between taste and smell does not apply to all combinations.

This has led to the concept of **congruence**. The pairing of the savoury MSG with the garlic odour of 2-methyl-3-furyl disulfide is congruent and reinforcing, as is the sweet taste of saccharin and the almond odour of benzaldehyde; however, MSG and benzaldehyde are incongruent and do not reinforce.

Congruency is also observed for the perception of taste intensity at levels above detection threshold. For example, the perception of the sweetness of sucrose is enhanced by strawberry and lemon odours but not at all by the odours of ham, peanut butter or liniment.

❍ Of the strawberry and lemon odours, which do you think will have the greater sweetness-enhancing effect?

● Probably the strawberry odour, as it is more commonly associated with sweeter tastes, whereas lemon has a more sour association.

This is precisely what is observed, although the ability of a strawberry odour to enhance sweetness diminishes as the sucrose concentration is increased (Figure 4.4).

❍ Why do you think the ham odour doesn't enhance the perceived sweetness of sucrose?

● Presumably because it is an incongruent odour.

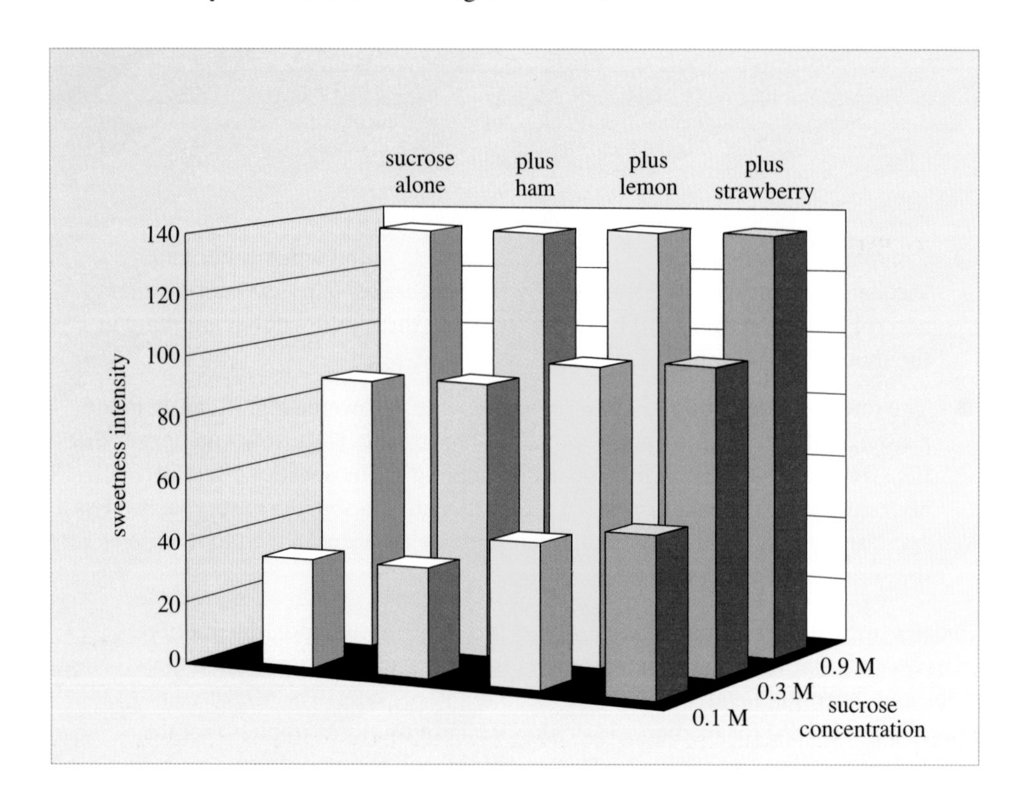

Figure 4.4 Sweetness enhancement of sucrose solutions by strawberry, lemon and ham odours.

Indeed, when the **hedonic perception** (that is, whether it is pleasant and we are attracted to it or unpleasant and we try to avoid it) of the ham/sucrose combination is examined, the ham odour reduces the pleasantness of sucrose solutions by the same magnitude as it reduces the perceived pleasantness of water. It would appear, therefore, that incongruent stimuli interact subtractively as far as hedonic perception is concerned.

We have already discovered that smell and taste involve separate peripheral detection systems. Consequently, integration of the signals from these two systems to generate the perception of flavour must be based somewhere in the central nervous system. Indeed, crossmodal interactions between any two sensory systems, as is seen to exist for taste and smell, demands that there is some higher-level point to which the two modes converge and integrate their inputs. For the perception of flavour, there are three potential sites where this might happen: the insular cortex, the orbitofrontal cortex and the amygdala (Figure 4.5).

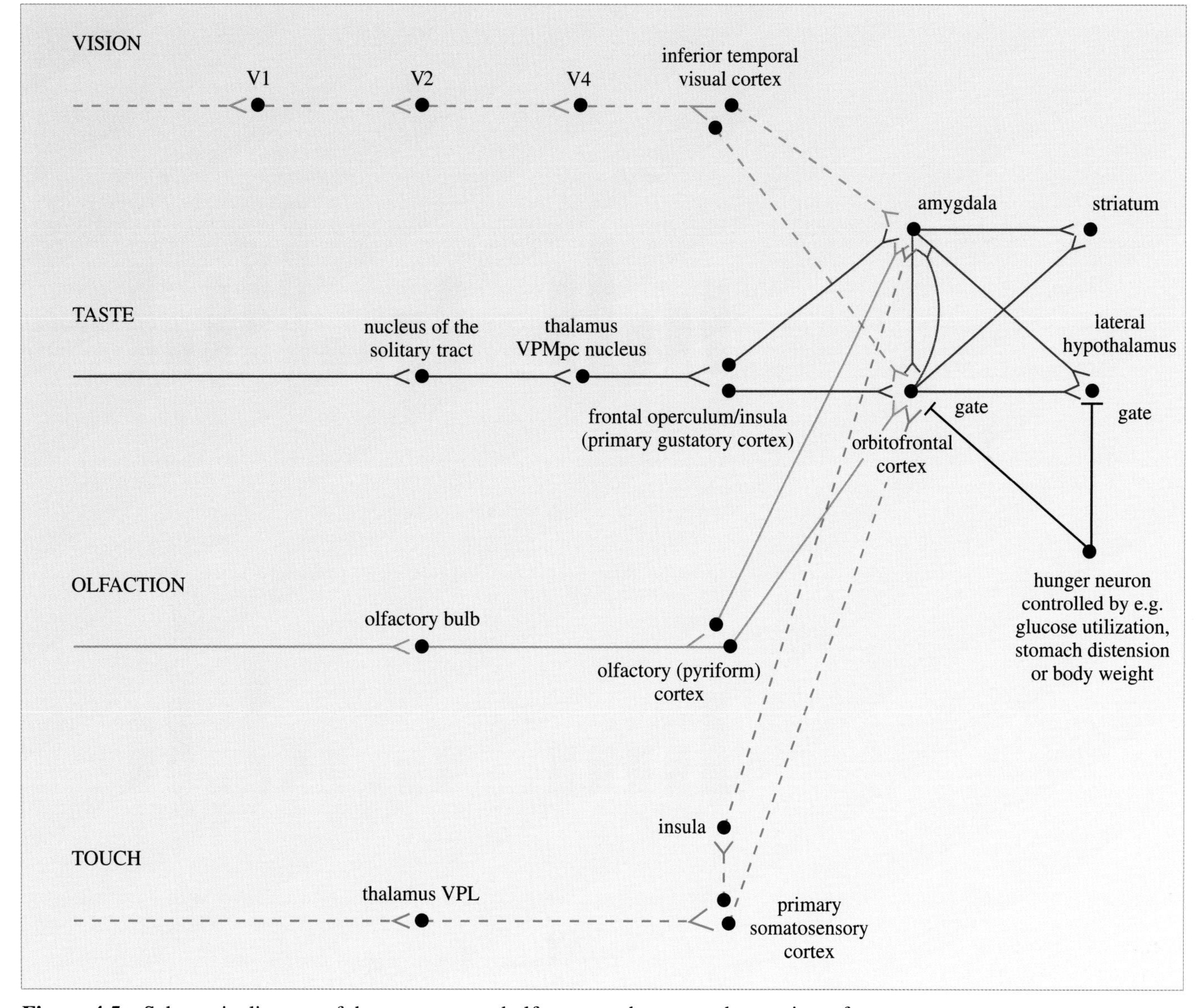

Figure 4.5 Schematic diagram of the gustatory and olfactory pathways to show points of convergence.

The observation of congruency suggests that integration is not a general gustatory–olfactory phenomenon, but one that arises from repeated exposure to particular tastant–odorant combinations. Indeed, when an unfamiliar odour is rated for its sweetness, that rating increases after it has been repeatedly paired with the sweet taste of sucrose. Similarly, its sourness rating increases when it has been repeatedly paired with citric acid. On the basis of such observations it has been suggested that the amygdala is the site for such processing, though there is, as yet, no proof of this. However, there is evidence that the orbitofrontal cortex can fulfil the role of 'flavour processor'. This comes from the work of the British scientist Edmund Rolls, who has examined how the neurons in various parts of the gustatory and olfactory pathway respond to odorants and tastants. Single neurons in the orbitofrontal cortex can respond to gustatory and olfactory inputs either independently or they can respond to both sensory modalities. About 2 per cent of the neurons of the orbitofrontal cortex respond to gustatory stimuli only, 1 per cent to olfactory stimuli, and just below 1 per cent to both types of stimuli. The response profiles of two cells that respond to both types of stimuli are shown in Figure 4.6.

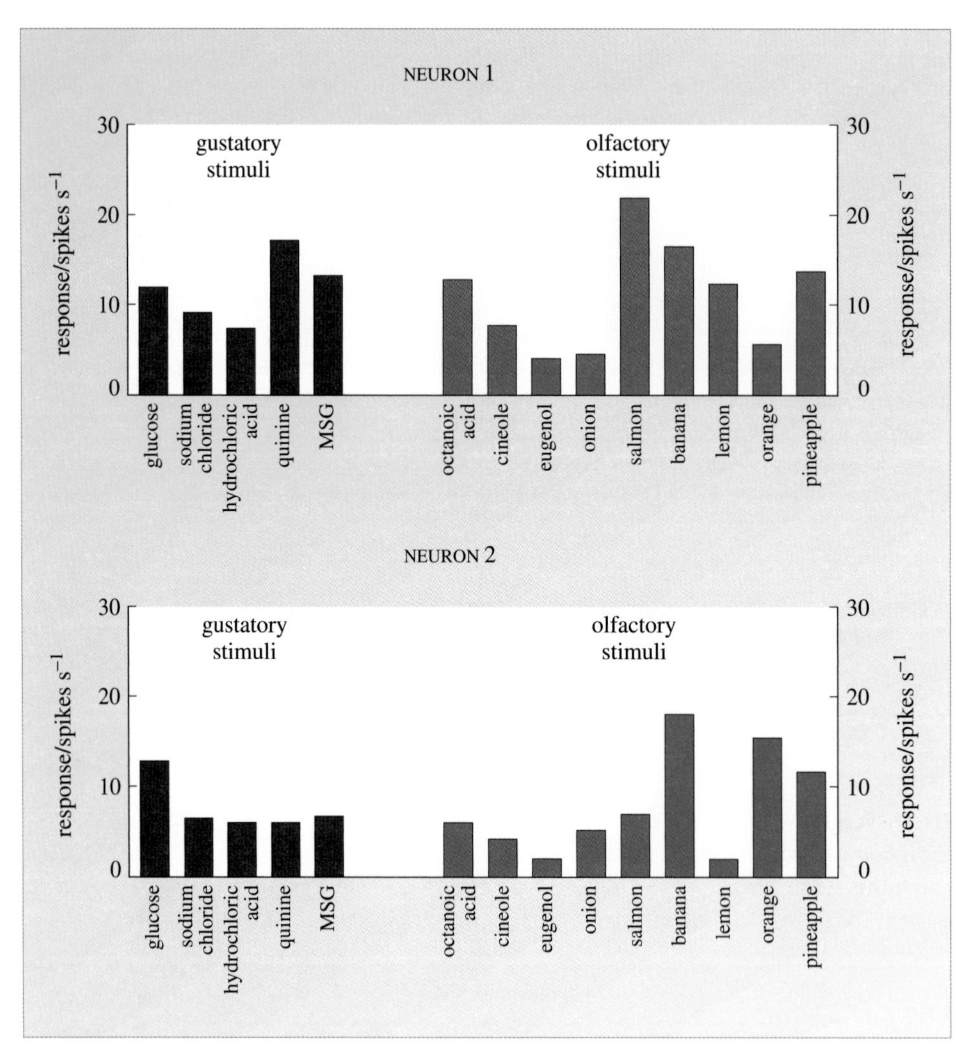

Figure 4.6 Response profiles of two orbitofrontal cortex neurons that respond both to olfactory and gustatory stimuli.

❍ Compare the profile of neuron 1 to olfactory and gustatory stimuli with that for neuron 2 to the same stimuli.

● Neuron 2 appears to be more narrowly tuned, both for gustatory ('sweet best') and olfactory ('fruit best') stimuli. Neuron 1 is more broadly tuned across both senses, responding best to bitter and umami tastes and the savoury salmon and rancid octanoic acid odours. Neuron 2 clearly exhibits congruence between sweet taste and fruit-related odours, as does neuron 1 to averse stimuli.

Since none of the cells in the primary gustatory cortex, the region immediately prior to the orbitofrontal cortex in the gustatory pathway, were found to respond to olfactory stimuli, it is believed that the neurons of the orbitofrontal cortex are the first point to which smell and taste inputs converge. If this is so, then neuronal activity in the orbitofrontal cortex might be expected to reflect hedonic responses to food. When hungry, food tastes pleasant, when eaten to fulfilment it loses its pleasantness. This is paralleled by the neuronal activity: upon feeding glucose to a monkey until it can take no more, neurons of the solitary tract and of the primary taste cortex do not lose their responsiveness during feeding whereas the response of neurons in the orbitofrontal cortex falls to zero. It therefore appears that the orbitofrontal cortex is the anatomical site where the brain constructs a representation of flavour.

4.1 Summary of Section 4

Flavour is the perception that arises when the senses of smell and taste combine, together with tactile, temperature and trigeminal sensations. The intensity of sweet or umami tastes is enhanced in the presence of a congruent smell. Likewise, a congruent taste can lower the detection threshold of an odour. Incongruent combinations of smell and taste appear to lower hedonic perception. Congruency between smell and taste requires a central site for merging of the signals from the separate pathways, and the orbitofrontal cortex provides such a site.

Question 4.1

On a psychophysical sweetness test a strawberry-smelling odorant that is not tasted is perceived as being sweet. How can this be?

Question 4.2

When you have a cold, why don't you 'taste' your dinner as well as you do normally?

Question 4.3

In a psychophysical test using scales of 0–100, a tasteless caramel odour was assigned a 'sweetness' rating of 78 and a 'sourness' rating of 10. What effect do you think this odour might have on perceived sweetness when combined with sucrose, and on the perceived sourness when combined with citric acid.

Objectives for Block 6

Now that you have completed this block, you should be able to:

1. Define and use, or recognize definitions and applications of, each of the terms printed in **bold** in the text.
2. Recognize molecular properties associated with odorants and tastants. (*Questions 2.1 and 3.5*)
3. Describe in outline the neuronal pathways of olfaction and gustation. (*Questions 3.3, 3.4, 3.9 and 3.10*)
4. Describe and compare the mechanisms by which odorants and tastants bring about the production of a neuronal signal. (*Questions 2.2, 2.6, 2.9 and 3.8*)
5. Explain how odours and tastes are coded in the olfactory and gustatory systems, respectively. (*Questions 2.3, 2.6 and 2.7*)
6. Appreciate how olfactory adaptation occurs. (*Questions 2.5 and 2.10*)
7. Describe how olfactory discrimination is achieved, the methods by which it is examined and the limitations of the human olfactory system to discriminate odours. (*Questions 2.4, 2.7 and 2.8*)
8. Describe the main features of the five taste qualities: sweet, sour, salt, bitter and umami. (*Questions 3.1, 3.3, 3.7 and 3.8*)
9. Recognize and describe different types of taste interaction. (*Questions 3.2 and 3.6*)
10. Describe how the olfactory and gustatory senses interact to produce the perception of flavour. (*Questions 4.1, 4.2 and 4.3*)

Answers to questions

Question 2.1

$C_2H_4O_2$ has a relative molecular mass of 60 [(2 × 12) + (4 × 1) + (2 × 16)]. So both its mass and volatility (low boiling temperature) suggest it is a potential odorant.

C_2H_5NO has a relative molecular mass of 59. While its mass is sufficiently low its boiling temperature is quite high, so it may be an odorant though it is likely to be weak.

$C_3H_8O_3$ has a relative molecular mass of 92. However its boiling temperature is high so it is unlikely to be an odorant.

$C_5H_8O_2$ has a relative molecular mass of 100. Both its mass and boiling temperature suggest that it is a likely odorant.

Question 2.2

(i) The nine-carbon containing molecules in Figure 2.13 are nonanoic acid (entry 6) and nonanol (entry 11). Both molecules activate receptors S19 and S83 most strongly. While receptor S86 responds to nonanoic acid it remains unactivated by nonanol.

(ii) Receptors S1, S46, S79 and S86 are activated solely by carboxylic acids.

(iii) Receptors S3 and S25 are activated solely by alcohols.

(iv) To identify those receptors tuned to longer carbon chains we need to observe an increase in receptor response with increasing chain length, preferably for both series of compounds (acids and alcohols). The clearest examples of this trend are receptors S19 and S51. Receptors S18 and S83 may also respond to carbon chain length in both series, but the trend is less clear. Receptors S1 and S46 may also respond to carbon chain length, but only in the carboxylic acid series.

Question 2.3

(i)

Odorant	Receptor			
	A	B	C	D
1	s	w		vw
2	vs	m	vw	vw
3		vs		s
4	m	w	vw	s
5			vs	
6		s		vs

(ii) Activity is brought about by an odorant binding to the odorant receptor. When two different odorants are present the one that binds to the receptor the strongest binds preferentially. So, for a mixture of the odorants 2 and 3 the receptor pattern will be: A, vs (2 binds preferentially); B, vs (3 binds preferentially); C, vw (2 binds preferentially); D, s (3 binds preferentially).

Question 2.4

The molecular structures of butanol and octanol are more different (four CH_2 groups) than are those of heptanol and octanol (one CH_2 group). Consequently, butanol and octanol would be expected to activate combinations of receptors whose patterns differ more than the corresponding patterns produced by heptanol and octanol.

Question 2.5

Constant exposure of olfactory receptor neurons to odorants produces olfactory adaptation and desensitization. So the body odour sufferer's olfactory system adapts to the constant presence of the unpleasant odorants. However, the system of another person with whom they come into contact has not adapted to these odorants because they have not been exposed to them for any length of time. So the sufferer is unaware of the odour, but those nearby are.

Question 2.6

In a labelled-line understanding of olfaction, the olfactophore concept proved useful because a particular molecular feature, for example, the alcohol functionality, was thought to activate a receptor that brought about a particular odour perception, for example 'floral'. Now that olfaction is known to be coded combinatorially, the olfactophore concept remains valid because molecules possessing a particular bonding feature are likely to activate a common (though not exclusive) pattern of receptors, and it is this pattern that is converted into an odour perception – such as 'floral' or 'fruity'. So, building particular patterns into molecules is likely to afford them a desired odour attribute.

Question 2.7

The odour codes that will give rise to an odour being detected are:

abc, *abd*, *abe*, *acd*, *ace*, *ade*, *bcd*, *bce*, *bde*, *cde*, *abcd*, *abce*, *abde*, *acde*, *bcde* and *abcde*.

That means that 16 different odours can be detected. Of these, the following three are sufficiently different from each other to be discriminated:

abc, *ade* and *bcde*. (This is not a unique set, but the maximum number that can be discriminated is never more than three.)

Question 2.8

Four factors contribute to odour perception; odorant molecular structure, odorant concentration, adaptation and whether or not the odour is a mixture of odorants. Molecular structure determines the affinitiy of the odorant for individual olfactory receptors. Concentration affects perception because the pattern of glomeruli activity changes as the odorant concentration changes – the higher the concentration the more glomeruli that are activated. Since the pattern of the activated glomeruli codes the odour, the altered patterns can result in altered perceptions. Adaptation affects odour perception because prolonged exposure to an odorant results in reduced neuronal activity and loss of the perception. When odorants are presented as mixtures it is much more difficult to discern the individual components. If the mixture contains four or more odorants, identification of an individual odorant falls to the chance level.

Question 2.9

Different odorants display different affinities for their olfactory receptors. Some bind to olfactory receptors at very low concentrations, others only at much higher concentrations. Those odorants that bind at low concentrations therefore have a lower detection threshold.

Question 2.10

Repeated exposure results in adaptation and reduced olfaction perception. Since limes are related to lemons, many of the odorants present in lemon odour will be components of the lime odour. Consequently, the olfactory system will be adapted to these and the perception of lime odour significantly affected. However, the odour of peanut butter will contain few, if any, odorants in common with lemon odour. Consequently, the olfactory system will not be adapted to peanut butter odour and its perception will remain unaltered.

Question 3.1

Amiloride blocks the pathway that results in a sour taste. LiCl has a much more significant sour component, so in the presence of amiloride it will taste saltier but with a bitter component. KCl has little sour contribution to its perceived taste so it will remain dominantly bitter with a salty component.

Question 3.2

The mixture contains four components, so each component should be present at one quarter of its threshold detection level. These are, for quinine 0.35 μM, sucrose 160 μM, citric acid 17.5 μM and saccharin 2.5 μM.

Question 3.3

There is little evidence in Figure 3.33 to support the back of the tongue being the primary site for bitter taste. First, all regions of the tongue have roughly similar sensitivities to the bitter compounds quinine and urea (though the absolute sensitivities to each compound is different).

Second, of the compounds tested it is true that the back of the tongue is most sensitive to the bitter compound quinine. However, it is also true that the back of the tongue is more sensitive to the sour taste of citric acid and the sweet taste of sucrose than it is to the other bitter tastant urea. So, sensitivity to taste quality appears to be tastant dependent rather than taste-quality dependent.

Question 3.4

The chorda tympani branches of cranial nerve VII innervate the anterior two-thirds of the tongue. If these are cut then any taste sensation will be due to taste cells from the posterior one-third of the tongue and also from the soft palate. Figure 3.33 reveals these regions to respond to all taste qualities, so it is entirely consistent with the sense of taste being maintained. Furthermore, nerve fibres from the glossopharyngeal nerve are broadly tuned across all taste qualities, so it would be expected to convey a perception of taste across all these qualities.

Question 3.5

Salt is a water-soluble solid that has a high melting temperature. It is involatile, and to elicit an olfactory perception a molecule must be volatile so that it can enter the nasal epithelium. Its water solubility allows it to come into contact with taste cells on the tongue and soft palate.

Question 3.6

At the lower sodium chloride concentration sweetness enhancement (or possibly synergy) is observed, at the higher concentration sweetness suppression (or possibly masking) is present. Figure 3.29 reveals sodium chloride to be perceived as primarily sweet at 10^{-2} M, so the common perceptions of the two tastants enhance each other. At 10^{-1} M, sodium chloride is perceived mainly as salty, and this taste quality suppresses the sweetness of sucrose.

Question 3.7

The statement is incorrect. Figures 3.10 and 3.11 reveal that cells containing bitter taste receptors respond to more than one tastant. Similarly for sweet taste receptors, Figure 3.17 shows cells respond to a range of diverse sweet tastants.

Question 3.8

The statement is false. Sweet, bitter and umami tastes are activated by molecules binding to receptor proteins but sour involves organic and mineral acids bringing about a change in pH within the cell and salt involves the transport of ions through ion channels.

Question 3.9

Sweet and salt are the two taste qualities that bring about the highest activity in fibres of the chorda tympani and the lowest activity in fibres of the glossopharyngeal nerve (Figure 3.35). However, the activities that are observed for sweet and salt in the chorda tympani 'best' fibres are 12 and 9 spikes s^{-1}, respectively, and 9 and 10 spikes s^{-1} in the corresponding glossopharyngeal fibres. So, the chorda tympani and glossopharyngeal nerve respond to sweet and salt taste qualities with almost identical sensitivities.

Question 3.10

PROP taster status is directly linked to the number of fungiform papillae which are located at the front of the tongue: supertasters of the bitter tastant PROP have the highest density of fungiform papillae. So, PROP taster status is inconsistent with the idea that bitter is tasted at the back of the tongue.

Question 4.1

A strawberry odour and a sweet taste are congruent sensations. The odorant presumably activates neurons in the orbitofrontal cortex that normally receive the congruent inputs from both the olfactory and gustatory systems, the latter associated with a sweet taste. So, when these neurons are activated, the pathway by which the odour enhances a sweet taste is activated and a sweet taste perception is elicited.

Question 4.2

It is not taste that is being affected here but the sense of smell and how it interacts with taste to produce the perception of flavour. Secretions produced by colds block the nasal passages and restrict the access of odorants to the olfactory epithelium. Consequently, the population of orbitofrontal cortex cells that respond to smell (25 per cent) or smell and taste combined (25 per cent) has reduced inputs, and the perception of flavour is dramatically reduced.

Question 4.3

The caramel odour is congruent with a sweet taste but not a sour taste. Consequently, the odour would be expected to enhance the sweetness perceptions of sucrose and will either have no effect on the sourness of citric acid or it could diminish the sourness via a crossmodal interaction.

In reality, caramel odour does indeed enhance sweetness, and it also suppresses sourness.

BLOCK SEVEN

Integrating the Senses

Contents

1 Introduction 171

2 The role of attention 173

3 Hearing and attention 177

4 Sight and sound 179

5 Vision and the vestibular system 181

6 Failures of attention and integration 183

7 Perception and imagination 185

8 Conclusions 187

Objectives for Block 7 188

Answers to questions 189

Acknowledgements 190

Glossary for Blocks 5, 6 and 7 193

Index for Blocks 5, 6 and 7 201

Introduction

1

This final block of the course is concerned with integration. To some extent it will integrate material that you have studied so far, but principally it will consider the problem that the brain has of integrating sensory information. It may not be immediately obvious that there is a problem, but consider what the brain has to achieve in a rich sensory environment when one is bombarded with stimuli. An example that comes to mind is travelling on the London underground in the rush hour. Fortunately, that is not something that I have to endure very often, so when I do it makes an impression. The last time it was hot, and of course crowded, so olfaction was one of the senses stimulated, but let's draw a discreet veil over that and consider some of the other sensations. When we stopped at Piccadilly a large tourist squeezed in next to me; he wore a designer shirt, with a label in one corner, and was chatting to his companion. Now that I come to think about it, I must have been separately analysing, in my various sensory channels, all the sights and sounds, including the particular voice of the man beside me and his appearance. However, at the time I had a *single* experience, of a fellow passenger, whom I could feel pressed against me, who was producing speech sounds and whom I could see. I had no impression of having to assemble the experience from its constituents. How does the brain know which pieces of information, from the different senses, go together?

The issue of fusing information from different senses is only part of the problem, because we also have to integrate material from within single senses. Consider that designer label, which was in a corner and had blue lettering.

❍ Where in the sequence of analysis did these pieces of information become available?

● Basic colour information is available very early in analysis, when the outputs of the different cone types are compared.

  The first layer of the visual cortex is location specific, i.e. a given line will cause firing in different neurons, depending upon the line's location in the visual field. However, this layer is only able to recognize simple line segments.

  The neurons in later layers of the visual cortex are able to discriminate more complex shapes, but they fire for a given shape wherever it appears, over a wide area of visual field.

The above can be summarized by saying that the greater the complexity of information conveyed about a stimulus, the less is known, at that level of analysis, about its basic characteristics. The neural circuitry that recognized a trade name on a shirt would have responded wherever the word was printed, in whatever colour. Nevertheless, the position and colour information were somehow seamlessly combined with the perceived word.

We hope by now you are convinced that integration is an important issue. By the time you have completed this block you will have read about a variety of types of integration, both within and between senses. We shall now go on to consider examples of faulty integration and perception, and the mechanisms and regions of the brain that might be involved.

The role of attention

2

In the last section it was implied that my brain was processing a whole range of material as I travelled. The processing was assumed to be taking place in parallel, i.e. the various sources of information were being dealt with simultaneously, rather than having to be processed in sequence. Out of the plethora of stimulation I chose to attend to just a subset; presumably as a result I can remember details about the tourist, but nothing about all the other information to which I did not pay attention. Attention clearly has a role in influencing what we remember, but does it have additional importance? To answer that question we will first consider some simple examples of **parallel processing**.

Look at Figure 1, and try to get a feel for how long it takes you to find the blue letter. It will seem almost instantaneous, and certainly too quick to time with a stopwatch. Timing can be carried out in a computerized version of the test. Groups of letters are flashed on the screen, sometimes with an 'odd one out', and sometimes not. The task is to press a 'Yes' or 'No' key as quickly as possible, to indicate present/absent. It turns out that people can respond in less than a second, and the response times do not seem to depend upon how many letters are present (within reason): we can spot a blue letter among twenty red ones as quickly as between two. This implies parallel, rather than **serial processing**; we do not have to inspect each letter, one at a time, or the delay would increase with the number of letters inspected.

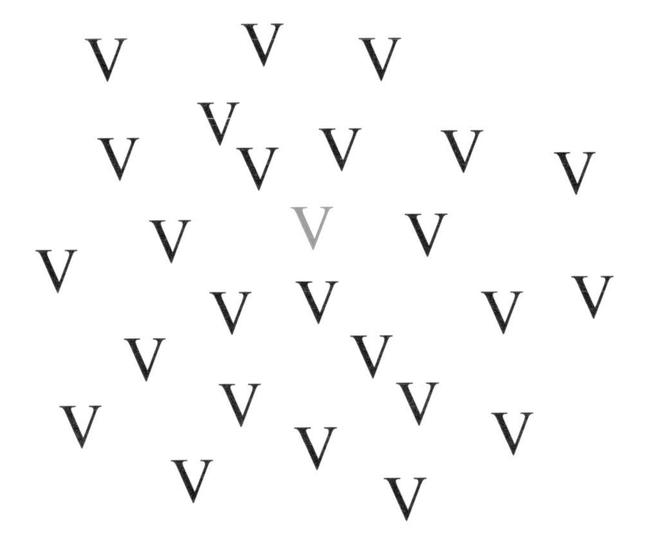

Figure 1 Find the blue letter.

You might feel that a colour contrast is trivially easy and proves little, but the demonstration has a little further to go yet. Have a look at Figure 2 (overleaf), and see how easy it is to locate the letter O. You will not be surprised to learn that this is another parallel process: the O can be found very quickly, whether there are few or many Vs. What happens if we now combine these two trivial, parallel tasks? Try to find the blue O in Figure 3 (overleaf).

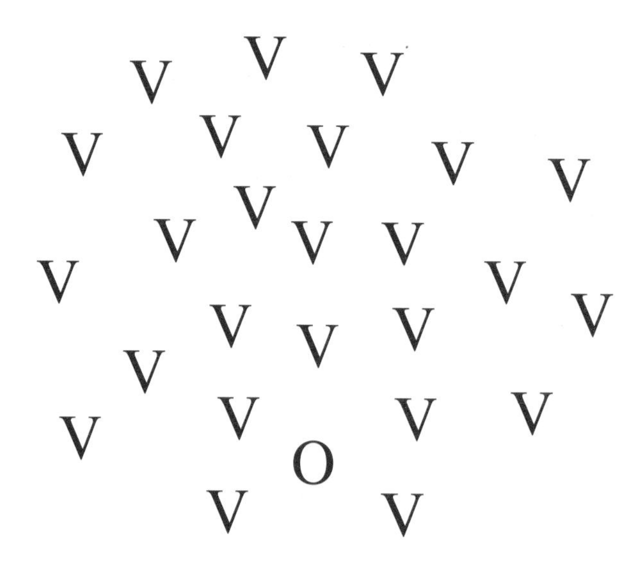

Figure 2 Find the round letter.

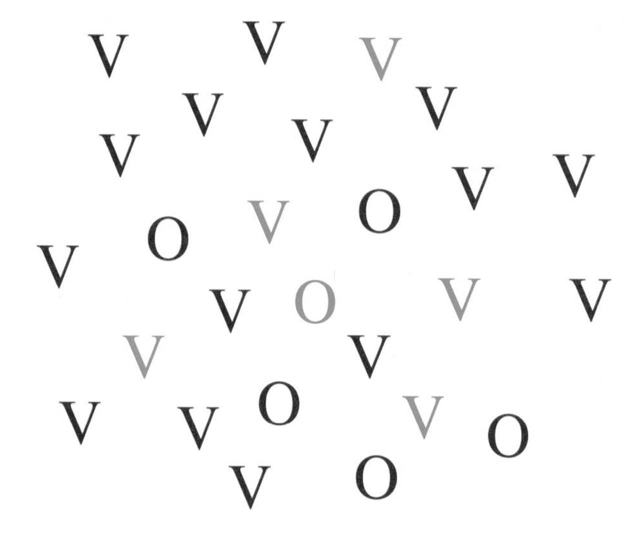

Figure 3 Find the blue O.

You may well have found that task just slightly harder, and research has shown that people do get slower as the number of distracting letters increases; in fact processing seems to become serial. This has been explained by suggesting that parallel processing is possible when only one distinguishing feature (such as colour, or shape) is to be found, but that serial searches are necessary when two or more features are to be found together. It is believed that focused attention is required when features are to be integrated in this way, the point of focus sweeping like a searchlight across the items in the scene, rather than illuminating them all simultaneously.

The effects of **feature integration** can also be demonstrated by presenting information very briefly using a **tachistoscope**. You may recall the description of extremely brief (tachistoscopic) displays given in Chapter 17, *The perception of words*, in the Reader. If people are shown a group of say ten letters, for about a twentieth of a second, they will probably remember only two or three of them. Nevertheless, if encouraged to report any other letters that they think they might possibly have glimpsed, people can often name one or two more correctly. What they probably will not know is whereabouts those letters were, or what they looked like. Typically, a viewer might say, 'I think there could have been a Q, somewhere down in the bottom left and written in red.' Sure enough, there was a Q, but it was in the middle of the group and printed in green!

At first the above description sounds ridiculous – someone sees something without knowing what it looks like! However, recall that colours, positions and letters must all be encoded in different neural structures, which then have to be combined to give the complete perception. A person's 'Q detector' (there must be a group of neurons that respond when Q is seen, so we can call that the Q detector) has to become active, for the attention process to unite the information with the other characteristics. When the display is tachistoscopic, the relatively slow, serial, attention mechanism is unable to bind many of the stimulus composites together before the display is gone again, but that does not mean that other letter detectors were not activated. Importantly, we appear to have great difficulty in becoming aware of what might be called the *meaning* of a stimulus, unless we have also united this with its specific physical characteristics. In our example, the person shown the letters would not have mentioned the Q unless encouraged to report the vaguest impression; the Q concept had been activated, but without associated physical details it would not have felt like 'really seeing it'.

We referred above to the meaning of a stimulus; a single letter, such as Q, does not have much meaning, but a word or picture certainly does. Research in this area has produced conflicting results, but the following appears to be possible. The 'moving' image seen at a cinema actually consists of a rapid sequence of still 'frames', each shown for one twenty-fourth of a second. The film can be cut, and another frame inserted in the sequence; it might, for example, be the name or picture of a brand of soft drink. When the film is shown, the fleeting display of the additional frame is too brief for anyone to see it consciously because the physical details could not be combined with the meaning in the time available. However, that meaning does seem to be extracted from the display. We know that it is, because sales of the drink go up in the interval! This is called **subliminal** advertising, and for obvious reasons it is illegal.

3 Hearing and attention

This brief section is intended to demonstrate that the effects of attention are not confined to vision. It will also give you the opportunity for some personal experience of the phenomena.

In hearing, as in vision, if attention is not applied to the stimuli, the recipient will remain more or less ignorant of them. This can be demonstrated by means of **dichotic listening**, i.e. listening to different things in the two ears. To ensure that you really do devote all your attention to one ear, you should 'shadow' everything you hear spoken in that ear. To **shadow** in this context means to follow closely behind, repeating everything you hear.

Activity 3.1 Dichotic listening

You should now undertake this CD-ROM activity. Further instructions are given in the Block 7 *Study File*.

Do not read on until you have carried out Activity 3.1.

Do you remember anything about what was heard by the unattended ear during the first part of the activity? The sex of the second speaker, for example, or the language? Many people do not. In fact it was a woman speaker, reading a piece in Spanish by the Chilean poet Pedro Neruda.

Assuming that you were unable to remember much about the message that you did not shadow, it would be easy to conclude that, by not attending, you did not extract any information. However, that was shown not to be the case by the second part of the activity.

In that, you probably found yourself moving to the other ear at one point, because the messages swapped ears part way through. A sentence that began in the shadowed ear finished in the other, and *vice versa*. Notice that you would not know about the shift, unless all along you had been extracting meaning from the unattended source.

4 Sight and sound

When trying to listen to someone in noisy surroundings, you may have found yourself paying close attention to their lips. Although most of us would not claim any lip-reading skills, it seems that we do understand speech better when we can see the lips that are producing it. Just how powerful this effect is can be demonstrated by deliberately making voice and lips mismatch, i.e. the lips move as if producing a different sound from the one that is actually heard. The perception that results is of a strange compromise, which matches neither voice nor lips. The phenomenon is called the **McGurk effect**, after the psychologist who first investigated it.

Activity 4.1 The McGurk effect

You can see a demonstration of the McGurk effect on the CD-ROM. As well as watching and listening simultaneously, you should try watching without sound, or listening without looking.

You should now undertake this CD-ROM activity. Further details are given in the Block 7 *Study File*.

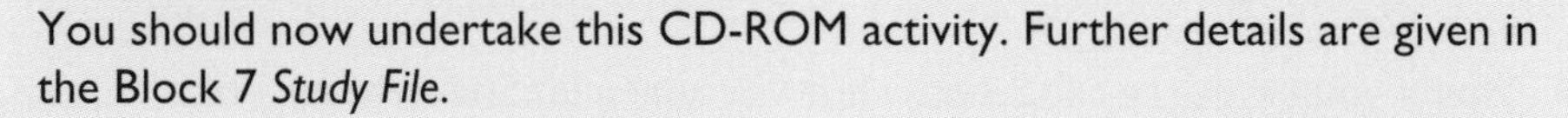

The McGurk effect demonstrates how our brains automatically fuse information (this time from two different senses) to produce a single perception, even when the result is not like either of the contributing stimuli. In its attempts to integrate, the brain will override perfectly good direction information. For example, when we watch someone speaking on the television we are not plagued by the feeling that the voice is coming from the wrong place: it seems to come from the moving lips. Obviously no sound can really come from a glass screen! As an exercise you might try covering your TV with a sheet and getting people to point to whereabouts the sound seems to be emanating. They are not likely to point to the middle of the screen; they will indicate the region of the loudspeaker (unless there are two, producing stereo). This indicates that we are well able to sense the sound direction, but the brain conveniently ignores that information, in the interests of sensory integration.

The TV exercise shows the influence of vision upon hearing, but the reverse can also take place. Before we give an example, it is necessary to explain some distinctions between different sorts of attention deployment. The eyes and ears have complementary strengths and weaknesses. Vision has the advantage that the eyes (and head) can be turned, so as to ignore irrelevant, distracting material, but the penalty is the possibility of missing important information. In contrast, the ears cannot be oriented significantly; in principle, all auditory information is permanently available. However, this does mean that the listener can be distracted by irrelevant sounds. For a person who is deaf in one ear, or who wears a single hearing aid, following speech in a noisy environment is very difficult.

❍ Why do people with both ears functioning manage to cope in noisy surroundings?

● The answer lies in the fact that the brain compares information from the two ears, to establish the directions of the various sounds (as described in Block 3).

If sounds can be localized, then attention can be directed towards the signal of interest; it is almost like an internal equivalent of turning the eyes. This 'internal' directing of attention is said to be **covert**, i.e. hidden, because an outside observer would not be able to tell which sound source was the object of attention. This is in contrast with judging the focus of visual attention, where the point of interest is generally clear, and is said to be **overt**, that is open or unconcealed. However, although this ear–eye distinction seems clear, we can actually be covert in our direction of visual attention. To take the underground as an example again, it is quite possible to carry out some examination of one's fellow passengers, while seeming to be studying the route map!

In addition to the covert–overt distinction, there is also one to be made between two causes of attention deployment. If we decide to examine someone out of the corner of our eye, then it is our decision which directs the attention to that location; the shift is said to be **endogenous**, i.e. internally originating. However, if while reading, for example, we glimpse a sudden movement from the corner of our eye, then we will have our attention pulled in that direction by the external event. The involuntary shift is termed **exogenous**, meaning externally caused.

❍ Would you expect an exogenous shift in visual attention to be covert?

● No, generally, if our attention is caught by something external, we will turn to look at it; this is clear for all to see, and therefore is overt.

However, there can be times when there is something of an exogenous–endogenous battle, and the result may be a covert shift. The example we shall give brings us back to the issue of whether hearing can have an impact upon vision, and it concerns the use of mobile telephones in cars.

Many drivers have abandoned the dangerous practice of clutching a 'phone in one hand, changing gear with the other, and holding the steering wheel with their knees; they now use hands-free sets. These attachments consist of a cable, with a microphone and single earpiece. The sound is inevitably delivered to one ear, so, if the driver wants to listen to what the caller is saying, it is necessary to make an endogenous shift of auditory attention to one side. Meanwhile, it is to be hoped that visual attention is directed forwards, and that is very likely the direction in which the eyes will be pointing. However, there is evidence that the sideways shifted auditory attention exerts an effect on vision, causing a covert shift. Without full attention being deployed to the front, response times to the traffic ahead are likely to be longer, with the obvious result of an increased accident risk.

Issues such as the above are discussed in Chapter 27 of the Reader, *Multisensory integration, attention and perception* by Charles Spence. You should read that before proceeding with the next section.

Vision and the vestibular system

5

Reader Chapter 27 looks at the interaction between vision and most of the other senses, except the sense of balance. Even without considering formal research, one would expect sight and balance to work together, since both can inform us about our orientation or movement. The importance of vision to maintaining balance can be demonstrated by getting someone to stand on one leg with their eyes closed. Although this is an easy task when there is vision to help signal movement, it becomes quite difficult in the absence of that cue. Nevertheless, we are able to walk about in the dark, so the vestibular system is clearly quite effective in helping us to maintain our balance. People who suffer from Menière's disease, which damages the vestibular system and produces episodes of giddiness, tend eventually to lose vestibular function. At this stage they become free from the vertigo, but have to rely upon vision alone for balance. This proves perfectly possible, but does mean that, unlike others, the sufferer cannot walk in the dark.

As with other sensory combinations, if there is a mismatch between the signals from the eyes and balance organs, then one or the other tends to dominate. Very often it is vision that determines the outcome, an effect which seems particularly strong in young children. This can be demonstrated, using three sides of a little room, rather like part of a 'Wendy House', but with no floor. The structure is either mounted on rollers, or suspended on cords, so that it can be moved. A toddler stands inside, facing a wall, with another wall on either side, so that the room fills the child's field of view. The room is then moved abruptly towards the child, who immediately falls backwards. This occurs, because our environment, such as walls, is generally static; if a wall seems to be approaching it is most likely because we are falling towards it. The toddler's brain makes that deduction, and initiates corrective movements, to counteract the (illusory) forward fall. The unnecessary corrections result in the backward tumble. Adults are more resistant to this effect, although audiences at IMAX® cinemas (which, like the three-sided room, fill the field of view) can be seen to sway (or stagger, if standing), when there are fast moving sequences, such as roller-coaster rides.

In other situations it is the vestibular system that can dominate the response, sometimes with the potential for disastrous results. An example of this occurs when fighter pilots take off from an aircraft carrier. Because the ship is much shorter than a conventional runway, there is insufficient distance for the 'plane to reach take-off speed under its own power alone. To achieve the necessary speed in time, a catapult system is used, which hurtles the aircraft along the deck with far greater than normal take-off acceleration. You will probably have experienced the tendency for the head to roll back, when sitting in a vehicle which is accelerating forward; a similar, but more powerful effect acts upon the pilot's otoliths (see Reader Chapter 7, *The vestibular system*, if you have forgotten their role in the vestibular system). The backward movement of the otoliths would normally occur only if a person tilted backwards, such as when the dentist tilts the chair. The conclusion 'I am tilting' is the interpretation that a pilot's brain makes, and in spite of the fact that his eyes will be indicating that the aircraft is still horizontal, he will feel that it is pointing up at a steep angle. On its own, this misperception might not matter too much, because things would quickly return to normal, once the 'plane was airborne. However, pilots know that to climb too steeply at take-off will cause the aircraft to stall and crash. After years of training, the automatic response to the sensation of tilting is to push the joystick forward, to reduce the rate of climb. As the 'plane was not in fact climbing, the result of this 'correction' is to direct it downwards and it flies into the sea! Because the illusion of climbing is so hard to resist, carrier pilots are instructed to take off with their hands in their lap, not on the joystick.

Another cause of illusory motion, this time involving the semicircular canals, is over-indulgence in alcohol. If one drinks heavily, an appreciable concentration of alcohol builds up in the blood stream and other body fluids. However, it is slow to reach the fluid of the vestibular system, as that is buried in the bone of the skull. This results in a concentration difference between the inside and outside of the semicircular canals, and that causes a small pressure difference, just enough to distort the canal membrane. Normally, that distortion would have been the result of rotating movements of the head, so the brain draws the obvious conclusion – that the sufferer is spinning! Keeping one's eyes open, in an attempt to convince oneself that everything is stationary, only adds to the problem. The vestibular system signals movement, while the visual system produces contradictory information; when there is a mismatch between the two, the result is frequently a sensation of nausea, as with sea-sickness.

These topics are explored further in Chapter 28 of the Reader, *Vision and the vestibular system* by Rollin Stott; you should read this before moving on to the following section.

Failures of attention and integration

6

By now it should be clear that attention is required to assemble and integrate sensory information, and that without integration the stimuli can fail to reach conscious awareness. In Section 2, some examples were given, showing how the decision to attend to one piece of information (e.g. a message in one ear) will result in little being known about other, neglected material. The decision to attend or ignore is one that we are constantly making. Sometimes we become so engrossed with one set of stimuli (the print in a book, perhaps), that the 'ignoring' seems total, and people may have difficulty in attracting our attention. Generally, however, changes in seemingly unmonitored information sources do succeed in capturing attention from the task in hand.

❍ How might unattended stimuli be able to attract attention, if they are not being processed in the first place?

● Part of the answer must lie in the fact that material *is* processed, even if we never become aware of it. Nevertheless, if we have no conscious awareness of a stimulus, it is not clear how we could decide to attend to it; endogenous attention shifts would not be expected. That means exogenous processes are implicated, and that they must take place more or less automatically.

Since complex stimulus integration does not occur in the absence of attention, exogenous attention shifts are brought about by simple stimulus characteristics, such as seeing a sudden movement, hearing a change in the ambient sound, or feeling a touch. In line with this conclusion, **sub-cortical** areas have been shown to be involved in more reflex attentive processes, while the cortex (particularly in the parietal lobes) deals with voluntary attention. (Refer back to Reader Chapter 27 if you do not recall these details.)

People suffering brain injury, particularly stroke victims, frequently sustain cortical damage; it is sometimes bilateral, but more usually in a single hemisphere. Those unfortunate enough to receive injuries to the parietal lobes of *both* hemispheres may display a condition known as **Balint's syndrome**. The patient shows a fixity of gaze, and seems unable to sweep their attention around. Once they do establish attention to an object, patients find it very hard to shift, with bizarre implications for the analysis and integration of other nearby stimuli. For example, a sufferer may say, 'When I see your spectacles, I can no longer see your face'.

Fortunately, Balint's syndrome is rare, but unilateral cortical damage is common in stroke victims, and this may have an impact upon attention and hence perception. In a common 'rough and ready' test of stroke, the investigator holds his or her hands up, on either side of the face, while getting the patient to look at his/her nose, i.e. to direct the gaze between the two hands. The investigator then wiggles a finger on one hand or the other, and asks the patient whether s/he can see it; if there is no damage to the visual cortex, this should present no problem, whichever side is moved. However, if a finger from each hand is moved simultaneously the patient will typically only report seeing movement on one side. The side that is seen is ipsilateral (i.e. same side) to the site of the brain injury, while the contralateral side is missed.

❍ Why do you suppose information is missed on the side contralateral to the injury?

- The majority of sensory pathways cross. For example, the left visual field is processed in the right visual cortex. The attention-related structures are located on the same side as the sensory regions that they direct. A person with damage to the right parietal lobe will have difficulty in directing attention to events on the left.

An extreme form of injury-induced, 'one sided' attention occurs in a condition known as **sensory neglect**. A patient may ignore a visitor who stands and speaks from the unattended side, and might even eat food from just one side of the plate. When asked to draw a clock face, a neglect patient managed a complete circle, but then populated it with numbers only from 12 to 6. It would be tempting to assume that a person so profoundly ignorant of information from one side must have injury to primary sensory pathways, but the problem persists, even when the senses are not directly involved. One patient was invited to imagine standing in his home city square, facing the cathedral, as he must have done many times before the stroke. He was asked what he could remember. In response he described the buildings on the left of the square but omitted those on the right. He was then asked to imagine looking out on the square from the cathedral steps, i.e. looking in the other direction. Again, his description was of buildings to the left, rather than the right, but the change of imaginary viewpoint meant that previously unmentioned monuments were now described, while the others were omitted! Clearly, his memory was intact but access to it was distorted by the attention problem. This problem lies in the endogenous system and ploys such as producing movement on the unattended side (to activate the exogenous system) can sometimes ameliorate the patient's difficulties.

7 Perception and imagination

The account of the patient imagining the view in the cathedral square shows the close link between imagined and real perceptions. This also serves as a reminder that there is another type of attention shift that we can make – between our 'inside world' and the outside world of reality. It can become very internally directed, as when a person becomes absorbed in a book (an example given at the start of the previous section). In some cases it can seem that all the attentive processes are involved with constructing the world described in the novel, and that inner construction takes on something approaching reality. People vary in the extent to which they have that kind of experience; those more inclined to do so are also more likely to respond to hypnosis. A possible explanation of hypnotic effects is that the hypnotized individual ceases to engage in external reality monitoring, and directs attention almost exclusively to internally generated imaginings. The region of brain that appears to be involved is in the frontal lobes. Earlier in this block, the parietal lobes were described as the locus of conscious direction of attention, but their role appears to be to direct attention on the basis of stimulus type and location. A more 'executive' role is taken by the frontal region, including the allocation of attention between internal and external perceiving.

The use of hypnosis as a means of exploring perception is described in Chapter 29 of the Reader, *Perceiving, misperceiving and hypnotic hallucinations* by Peter Naish. You should read this before tackling the concluding section.

8 Conclusions

In this final block of the course you have been confronted with the intricacy and delicacy of perception. For animals much simpler than ourselves, perception can be a rather robust and mechanistic process, where a given type of stimulus can be relied upon to elicit a given behaviour. For us, however, stimuli that are present may be missed, while those that are absent may be perceived, and many others will give rise to 'compromise' experiences. Our perceptions are not a predictable consequence of sensory stimulation, but are dramatically influenced by the processes of attention.

Integration is essential to a rich sensory experience. Few stimulus sources in our environment give rise to simple, single-sense stimulation, but our experience of them is of a unitary complex. Moreover, there is a synergy between the senses, such that activity in one can prime the corresponding structures of another. Reader Chapter 27 gives the example of activation in the auditory cortex in response to the *visual* experience of seeing lips move. The description of signal detection theory in Chapter 29 gives an account of how priming can occur and how it can even result in misperceptions.

The ability to recall previous perceptions in the form of mental 'images' (not just visual, but the other sensations too) shows that integration takes place during memory too, and the descriptions of visual neglect show that attention is required, even to deal with internally generated perceptions. Fortunately, we are generally able to distinguish between perceptions with external origins, and those that are internally generated. However, as a last thought, a good case could be made for claiming that *all* perceptions are internally generated.

Question 8.1

Which sense can be said to 'dominate' in the majority of situations?

Question 8.2

Why is a person less likely to feel seasick if they watch the horizon, rather than the interior of the ship?

Question 8.3

When someone is 'lost' in a book, in which direction might their signal detection criteria shift?

Question 8.4

Which brain regions have been identified as being involved in producing unified sensations from the stimulation of two or more senses?

Objectives for Block 7

Now that you have completed this block, you should be able to:

1. Define and use the terms printed in **bold** in the text, and recognize their definitions and applications.
2. Explain the importance of the integration of information, both from within and between the senses. (*Questions 8.1 and 8.2*)
3. Understand the role of attention in the formation of conscious awareness of stimuli. (*Question 8.3*)
4. Describe and cite examples of the influence of one sensory modality upon another. (*Questions 8.1, 8.2 and 8.3*)
5. Give examples of ways in which stimuli may be missed, or misperceived. (*Questions 8.1 and 8.3*)
6. Describe the effects of mismatches between information from different sensory sources. (*Question 8.2*)
7. Name examples of brain regions that have been identified as being involved in producing unified sensations from composite stimuli. (*Question 8.4*)
8. Recognize the relevance of information from this block to issues addressed earlier in the course.

Answers to questions

Question 8.1

In many instances, the sense of vision seems to exert the greatest influence upon the final interpretation of a multi-sensory stimulus. Reader Chapter 27 discusses visual dominance within multisensory integration in more detail.

Question 8.2

Nausea results from mismatch between movement information derived via the eyes and from the vestibular system. Inside a ship, the surroundings look stationary, as they move up and down with the passenger. The balance organs, in contrast, signal the actual movement, so sickness can result. If the passenger views the horizon, then the eyes too will signal that the ship is going up and down, and a better match results. Reader Chapter 28 explains the sensory mismatch theory of motion sickness.

Question 8.3

If the reader becomes oblivious to outside distractions, presumably the criteria for such signals have been raised, so that they will not reach consciousness. On the other hand, if parts of the book begin to feel almost real, it is possible that, in some people, the criteria relating to the scenes described become lowered sufficiently to give rise to appropriate sensations. A more detailed description of the signal detection theory is given in Reader Chapter 29.

Question 8.4

The parietal lobes are responsible for voluntary attention; the frontal lobes deal with allocating attention between internal and external perceiving. As such, they both have a unifying role. Sub-cortical areas are involved in reflex attentive processes.

Acknowledgements

Grateful acknowledgement is made to the following sources for permission to reproduce material in this book:

Cover

Coloured scanning electron micrograph of the surface of the nasal cavity. Copyright © Susumu Nishinaga/Science Photo Library.

Block 5

Figures

Figure 2.4: Knibestol, M. (1973) 'Stimulus-response functions of rapidly adapting mechanoreceptors in the human glabrous skin area', *Journal of Physiology*, **232**, pp. 427–452, Cambridge University Press; *Figure 2.6*: Schmidt, R. F. (ed.) (1978) *Fundamentals of Sensory Physiology*. Copyright © 1978, Springer Verlag; *Figure 2.7*: Kandel, E. R. *et al.* (2000) Chapter 22, 'The bodily senses', *Principles of Neural Science*, The McGraw-Hill Companies Inc; *Figure 2.11*: Johansson, R. S. and Vallbo, Å. B. (1983) 'Tactile sensory coding in the glabrous skin of the human hand', *Trends in Neuroscience*, **6**, pp. 27–32, Elsevier Science; *Figure 3.2*: Hensel, H. and Kenshalo, D. R. (1969) 'Warm receptors in the nasal region of cats', *Journal of Physiology*, **204**, pp. 99–112, Cambridge University Press; *Figure 3.4*: BSIP Astier/Science Photo Library; *Figure 3.5*: Stanley B. Burns, MD and The Burns Archive, New York/Science Photo Library; *Figure 4.4b*: Nicholl, J. G. *et al.* (2001) *From Neuron to Brain*, 4th edition. Copyright © 2001 Sinuaer Associates, Inc; *Figure 5.7*: Anderson, D. J. (ed.) (1980) 'Touch and pain – facts and concepts, old and new', *Physiology: Past, Present and Future*, Pergamon Press Limited; *Figure 5.8*: Lynn, B. (1979) 'The heat sensitization of polymodal nociceptors in the rabbit and its independence of the local blood flow', *Journal of Physiology*, **287**, pp. 493–507, Cambridge University Press; *Figure 5.13*: Solonen, K. A. (1962) 'The phantom phenomenon in amputated Finnish war veterans', *Acta Orthopaedica Scandinavica,* Supplementum No 54, Blackwell Publishers Limited.

Block 6

Figures

Figure 2.8: Singer, M. S. (2000) 'Analysis of the molecular basis for octanal interactions in the expressed rat 17 olfactory receptor', *Chemical Senses*, **25**, pp. 155–165, by permission of Oxford University Press; *Figures 2.12, 2.13*: Malnic, B. *et al.* (1999) 'Combinatorial receptor codes for odors', *Cell*, **96**, pp. 713–723. Copyright © 1999, with permission from Elsevier Science; *Figures 2.16, 2.17a*: Rubin, B. D. and Katz, L. C. (1999) 'Optical imaging of odorant representations in the mammalian olfactory bulb' *Neuron*, **23**, pp. 499–511. Copyright © 1999, with permission from Elsevier Science; *Figures 2.17b, 2.18*: Imamura, K. *et al.* (1992) 'Coding of odor molecules by mitral/tufted cells in rabbit olfactory bulb. I. Aliphatic compounds', *Journal of Neurophysiology*, **68**, pp. 1986–2002, The American Physiological Society; *Figure 2.20*: Friedrich, R. W. and Stopfer, M. (2001). 'Recent dynamics in olfactory population coding', *Current Opinion in Neurobiology*, **11**, pp. 468–474. Copyright © 2001, with permission from Elsevier Science;

Figures 2.22, 2.23: Sobel, N. *et al.* (2000) 'Time course of odorant-induced activation in the human primary olfactory cortex', *Journal of Neurophysiology*, **83**, pp. 537–551, The American Physiological Society; *Figure 2.24*: Qureshy, A. *et al.* (2000) 'Functional mapping of human brain in olfactory processing: a PET study', *Journal of Neurophysiology*, **84**, pp. 1656–1666, The American Physiological Society; *Figure 2.25*: Kajiya, K. *et al.* (2001), 'Molecular bases of odor discrimination: reconstitution of olfactory receptors that recognize overlapping sets of odorants', *Journal of Neuroscience*, **21**, pp. 6018–6025. Copyright © 2001 Society of Neuroscience; *Figure 2.26*: Rubin, B. D. and Katz, L. C. (1999) 'Optical imaging of odorant representations in the mammalian olfactory bulb' *Neuron*, **23**, pp. 499–511. Copyright © 1999, with permission from Elsevier Science; *Figures 2.27, 2.28, 2.29, 2.31, 2.33*: Zufall, F. and Leinders-Zufall, T. (2000) 'The cellular and molecular basis of odor adaptation', *Chemical Senses*, **25**, pp. 473–481, by permission of Oxford University Press; *Figure 2.38a*: Laska, M. and Hübener, F. (2001) 'Olfactory discrimination ability for homologous series of aliphatic ketones and acetic esters', *Behavioural Brain Research*, **119**, pp. 193–201. Copyright © 2001, with permission from Elsevier Science; *Figure 2.38b*: Laska, M. and Teubner, P. (1999), 'Olfactory discrimination ability for homologous series of aliphatic alcohols and aldehydes', *Chemical Senses*, **24**, pp. 263–270, by permission of Oxford University Press; *Figure 2.39*: Laska, M. *et al.* (2000) 'Olfactory discrimination ability for aliphatic odorants as a function of oxygen moiety', *Chemical Senses*, **25**, pp. 189–197, by permission of Oxford University Press; *Figure 2.45*: Jinks, A. and Laing, D. G. (2001) 'The analysis of odor mixtures by humans: evidence for a configurational process', *Physiology & Behavior*, **72**, pp. 51–63. Copyright © 2001, with permission from Elsevier Science; *Figure 2.46*: Livermore, A. and Laing, D. G. (1998) 'The influence of odor type on the discrimination and identification of odorants in multicomponent odor mixtures', *Physiology & Behavior*, **65**, pp. 311–320. Copyright © 1998, with permission from Elsevier Science; *Figure 3.2*: PDB ID: 1THU, published in Ko, T.-P. *et al.* (1994) 'Structures of 3 crystal forms of the sweet protein thaumatin', *Acta Crystallographica D*, **50**, pp. 813–825; *Figure 3.9*: courtesy of Professor Virginia Utermohlen, Cornell University; *Figure 3.11*: Caicedo, A. and Roper, S. D. (2001), 'Taste receptor cells that discriminate between bitter stimuli', *Science*, **291**, pp. 1557–1560, American Association for the Advancement of Science; *Figure 3.14*: Nofre, C. and Tinti, J.-M. (1996), 'Sweetness reception in man: the multipoint attachment theory', *Food Chemistry*, **56**, pp. 263–274. Copyright © 1996, with permission from Elsevier Science; *Figure 3.16*: Hoon, M. A. *et al.* (1999) 'Putative mammalian taste receptors: a class of taste-specific GPCRs with distinct topographic selectivity', *Cell*, **96**, pp. 541–551. Copyright © 1999, with permission from Elsevier Science; *Figure 3.17*: Nelson, G. *et al.* (2001) 'Mammalian sweet taste receptors', *Cell*, **106**, pp. 381–390. Copyright © 2001, with permission from Elsevier Science; *Figure 3.18*: Max, M. *et al.* (2001) 'Tas1r3, encoding a new candidate taste receptor, is allelic to the sweet responsiveness locus Sac', *Nature Genetics*, **28**, pp. 58–63. Copyright © 2001 Nature Publishing Group; *Figures 3.24, 3.25, 3.27, 3.28*: Lyall, V. *et al.* (2001) 'Decrease in rat taste receptor cell intracellular pH is the proximate stimulus in sour taste transduction', *American Journal of Physiology – Cell Physiology*, **281**, pp. C1005–C1013, The American Physiological Society; *Figure 3.29*: Ossebaard, C. A. and Smith, D. V. (1996) 'Amiloride suppresses the sourness of NaCl and LiCl', *Physiology & Behavior*, **60**, pp. 1317–1322. Copyright © 1996, with permission from Elsevier Science; *Figure 3.31*: O'Doherty, J. *et al.* (2001) 'Representation of pleasant and aversive

Glossary for Blocks 5, 6 and 7

active touch The type of intentional touch used to assess the properties of objects.

acute inflammatory response The series of local changes (pain, heat, redness and swelling) caused by the release of a variety of substances as a result of tissue damage.

adaptation The process by which the response to a stimulus is modified by extended or repeated exposure to the stimulus.

adequate stimulus The particular stimulus modality to which a receptor is responsive and which is transduced by the receptor.

affective visual information Visual stimulus or situation which produces an emotional response.

agonist A chemical substance capable of mimicking the physiological effect of a neurotransmitter by specifically interacting with its natural receptors.

allodynia Pain experienced when the subject is stimulated by a previously innocuous stimulus.

alpha-gamma co-activation (α-γ co-activation) The pattern of neuronal activity in muscles when both α and γ motor neurons fire causing simultaneous extrafusal muscle contraction (α motor neurons) and a stretching of the central portion of the spindle as a consequence of the contraction of their end portions (γ motor neurons).

amygdala An almond-shaped body located at the tip of the inferior horn of the lateral ventricle of the brain between the hypothalamus and the cerebral cortex. In general, the amygdala can be viewed as an area that is involved in regulation of awareness of behaviour.

analgesic A substance that reduces pain.

angular acceleration Acceleration leading to a change in the rate of rotation, e.g. in shaking or nodding the head.

anterior olfactory nucleus (AON) A sheet of cells just posterior to the olfactory bulb from which centrifugal fibres project to the olfactory bulb.

Balint's syndrome A condition resulting from bilateral brain damage, in which the sufferer experiences great difficulty in shifting attention and awareness from one component of a scene to another.

body image A perception of body shape, position and appearance, largely effected by visual feedback.

body schema An awareness of where all the body parts are at any one moment; the continuous, subconscious update of our postural model.

chemotopic map A map of the spatial distribution of taste cells and nerves in the oral cavity, different regions of which express varying sensitivities to the different taste qualities.

chiral recognition The ability to recognize mirror-image molecules as different.

chorda tympani A branch of cranial nerve VII that innervates the anterior two-thirds of the tongue including the fungiform papillae and the foliate papillae in the anterior folds.

cingulate cortex A region of cortex on the inward-facing surfaces of the frontal lobes.

circumvallate papillae Cylindrical structures surrounded by a trough that form a V-shaped line across the back of the tongue. The 9 or 10 circumvallate papillae together house almost half of the lingual taste buds, and the taste cells are innervated by the glossopharyngeal nerve.

combinatorial odour code The way in which an odour is coded by a particular pattern, or combination, of olfactory neurons being activated.

compliance Action made in accordance with a request.

conditioning A procedure for producing behavioural modification.

configurational hypothesis of olfaction An hypothesis which proposes that an odour perception is experienced following prior recognition of the odorant profile and the subsequent processing of the relationship between the components of the profile.

congruence In the context of smell and taste, the relatedness between the perception of a tastant and the odour of an odorant. Congruent odours and tastes reinforce the sensory perception.

covalent bond The bond formed by the sharing of two electrons between two atoms. This is the type of bond that holds atoms together in molecules.

covert (direction of attention) *See* covert orienting.

covert orienting Internal shift of attention (e.g. as when observing something out of the corner of the eye). *See also* overt orienting.

crossmodal Linking across more than one modality (i.e. sensory system), such as vision and hearing.

cyclo-oxygenase (COX) The enzyme responsible for controlling the formation of prostaglandins. There are two types called COX-1 and COX-2. COX-2 is activated as part of the acute inflammatory response to tissue damage.

dermatomes The area of skin whose sensory receptors are modified sensory neurons that lie in one single dorsal root ganglion.

dermis The inside layer of skin separating the epidermis from the underlying muscle, ligaments and bone; composed of connective tissue and elastic fibres floating in a semi-fluid called the ground substance, and containing many different cell types including tactile receptors and a rich blood vascular supply.

dichotic listening Listening simultaneously to two different messages, delivered by headphones, one message to each ear.

dorsal column nuclei The nuclei in the dorsal part of the medulla where the axons of the dorsal columns terminate. The neurons of the dorsal column nuclei relay touch information to the ventrobasal nucleus of the thalamus on the opposite side of the brain.

double blind clinical trials A protocol whereby neither the subjects nor the practitioners administering the treatment and the scientist analysing the results know which group received which treatment.

double dissociation The situation where two events that are usually causally linked become separate from one another. So event A is usually a necessary precursor of event B but if event A can occur without triggering event B and event B is manifest without event A having occurred, there is a double dissociation.

EC_{50} The concentration of a substance that produces an effect that is half the maximum response.

endogenous A process arising from within, generally implying that it is carried out with conscious intent. *See also* exogenous.

endorphins Naturally secreted substances, chemically like morphine, which produce analgesic effects.

enhancement The increase in intensity of a particular taste quality of one tastant due to the presence of another tastant.

enkephalins Peptides produced in the brain which bind to opiate receptors to block pain transmission; they are neuromodulators rather than neurotransmitters.

entorhinal cortex A caudally-placed region of five-layered cortex also found medial to the rhinal sulcus. This cortical area receives information from association cortical areas of all sensory systems and transmits information to the hippocampal formation.

epidermis The outer layer of skin; formed from epithelial cells and containing few, if any, tactile receptors.

event-related potential (ERP) A small voltage change, detected at the scalp, as a result of an event to which neurons have reacted.

exogenous A process triggered from the outside, generally involuntary. *See also* endogenous.

exteroceptive Responding to stimuli that are external to the animal.

facial nerve (VIIth cranial nerve) Also known as cranial nerve VII, the chorda tympani branch of this nerve innervates the taste buds in the anterior two-thirds of the tongue.

false alarm The detection of an apparent stimulation, which was in fact merely neural noise. *See also* miss.

feature integration The binding together of the various attributes of a stimulus, resulting in the conscious perception of a single, unified object.

filiform papillae Thread-like structures that cover the dorsal surface of the tongue; they have no gustatory function.

foliate papillae Flat-topped structures that have deep clefts between them, these papillae are found on the edge of the posterior one-third of the tongue at about the same level as the circumvallate papillae. Their appearance makes them look like leaves of a book (the word 'foliate' means 'leaf-like'). Foliate papillae house approximately one-quarter of the lingual taste buds.

forced-choice triangular test A test that involves differentiating between two stimuli by presenting three unknown options, two of which are identical, and asking the subject to choose the 'odd' stimulus.

fungiform papillae Raised structures on the surface and tip of the tongue that resemble miniature mushrooms. There are about 320 fungiform papillae on a human tongue and they house about a quarter of the lingual taste buds. Taste cells in fungiform papillae are innervated by the chorda tympani branch of the facial nerve.

fusimotor system The motor system of the muscle spindles.

gate-control theory of pain The theory, proposed by the British scientists Ronald Melzack and Patrick Wall in 1965, that there is a 'gate' at the level of the spinal cord that could regulate the effect of afferent axons from nociceptors signalling tissue damage. If the gate was closed, there would be no pain; if the gate was open, the afferents could excite the secondary neurons projecting up the spinal column leading to the perception of pain.

general senses Also known as the bodily senses, these are the senses for which receptors are widely distributed such as touch, pain, temperature, pressure and proprioception.

genome The total genetic material of an organism, comprising the genes contained in its chromosomes.

glabrous (skin) Hairless skin; one of three skin types, typified by skin on the palms and soles.

glomerulus (plural **glomeruli**) A spherical feature, approximately 150 μm in diameter, in the olfactory bulb in which the axons of olfactory neurons synapse with the primary dendrites of mitral cells. Each glomerulus receives axons from several thousand olfactory neurons, each of which express the same olfactory receptor, and receives dendrites from about 25 mitral cells. There are about 2000 glomeruli in the human olfactory bulb.

glossopharyngeal nerve (IXth cranial nerve) Otherwise known as cranial nerve IX, this nerve innervates the taste buds in the posterior one-third of the tongue.

Golgi tendon organ A proprioceptive sensory receptor in a tendon specialized for detecting tension within the tendon.

granule cells In the olfactory system, cells that interconnect mitral cells by forming synapses to their secondary dendrites and axons.

gustation The sense of taste.

hairy skin One of three skin types characterized by the presence of hair.

haptics Active touch.

hedonic perception Experiencing something as pleasurable or aversive.

hedonics The branch of psychology that deals with pleasurable and non-pleasurable states of consciousness.

hefting 'Throwing' an object up and down without allowing it to leave the hand.

histological study The microscopic study of the structure of tissues.

hydrogen bonding The bonding interaction that can occur between a hydrogen atom that is covalently bonded to oxygen or nitrogen and an oxygen or nitrogen atom elsewhere in the same molecule or in a different molecule. A hydrogen bond is about one-tenth as strong as a covalent bond.

hyperalgesia An increase in sensitivity to pain experienced in response to a given stimulus.

hypnotic induction The process of inducing hypnosis.

hypnotic susceptibility scale A series of tests, graded in difficulty, administered to a hypnotized individual, to determine the level of hypnosis that has been achieved.

hysterical paralysis A condition in which sufferers are unable to move a body part in spite of there being no identifiable physical problem.

inferior turbinate One of the three structures that arise from the lateral wall of the nasal cavity, the function of which is to create turbulent air flow and to clean, moisten and warm the inhaled air. The inferior turbinate is the lowest and largest of the three.

intrinsic signal imaging An optical technique that enables tissues to be monitored for changes in biochemical activity. The technique makes use of changes in optical properties due to, for example, blood volume.

ischemia A reduction of the blood supply to part of the body.

kinesthesis The sense that provides information about the movement of individual body parts. Also known as dynamic proprioception.

kinesthetic *See* kinesthesis.

labelled-line theory A coding hypothesis that implies a particular receptor neuron, or pathway of receptor neurons, is sensitive to and transmits information about a particular stimulus quality.

lateral olfactory tract The nerve formed by the bundle of axons projecting from the mitral cells. The tract projects primarily to the pyriform cortex and also to the anterior olfactory nuclei, the olfactory tubercle, the entorhinal cortex, and the amygdala.

linear acceleration A change in speed along a straight line; e.g. the rise and fall (i.e. a vertical straight line) of the head, when a person is running.

lingual gyrus Region of cortex on the ventral surface of the brain, hidden by the cerebellum; part of the primary visual cortex.

London forces The transient interactions between non-polar (hydrocarbon) regions of molecules. They are about one-hundredth the strength of a covalent bond.

long-lasting adaptation The adaptive response that can last for minutes even after very short exposure to an odorant.

masking In the context of smell and taste, a negative form of interaction that is stronger than suppression.

mast cells Large cells found in vertebrate connective tissue that produce histamine and serotonin in response to noxious stimuli.

McGurk effect The misperception of a spoken sound, resulting from listening to one speech sound while watching a speaker's lips forming a different sound.

mechanoreceptor A sensory receptor cell activated by mechanical forces such as pressure. They are involved in hearing, balance and touch sensations.

medial turbinate One of the three structures that arise from the nasal wall, the function of which is to create turbulent air flow and to clean, moisten and warm the inhaled air. The medial turbinate is a bony projection of the ethmoid bone, is the second largest of the three turbinates and lies between the inferior and superior turbinates.

Meissner's corpuscle A type of cutaneous sensory receptor cell found in glabrous skin that responds to light touch and vibration.

Merkel cells *See* Merkel's disc.

Merkel's disc A type of cutaneous sensory receptor cell that responds to light touch and pressure.

miss Failure consciously to detect a stimulus, when the resultant level of neural activity is insufficient to exceed an internally-set criterion. *See also* false alarm.

mitral cell A cell which has a primary dendrite that synapses with the axons of olfactory neurons in the glomeruli of the olfactory bulb. The axon of the mitral cell forms part of the lateral olfactory tract, a bundle of nerve fibres that lead to the olfactory centres in the brain.

modality-appropriateness The tendency of one modality (sense) to exert more influence than other modalities, in situations where the favoured sense is likely to convey more accurate information.

molecular receptive range The range of molecules, and hence molecular structures, to which an olfactory receptor is responsive.

monoaminergic pathways Neuronal pathways that use monoamines (e.g. serotonin and noradrenaline) as their neurotransmitters.

motion sickness A phenomenon in which the brain detects inconsistencies between vision and vestibular systems and this provokes a pattern of symptoms that include lethargy, light-headedness, sweating, nausea and vomiting.

mucocutaneous skin One of three skin types; it is mucus producing and lines the entrances to the body.

muscle receptors *See* muscle spindle.

muscle spindle A proprioceptive sensory receptor found in skeletal muscle; formed of specialized muscle fibres (connected in parallel to the ordinary muscle fibres) and sensory receptors; sensitive to muscle stretch.

nares Another name for the nostrils.

neural noise The random element of neural firing, which may be exhibited as firing in the absence of a signal, or variations in firing rate which are not correlated with signal intensity. This gives rise to misses and false alarms.

neurite An axon or dendrite. This is a term (like fibre) that can be used for any neuronal process. It is especially useful when it is not clear whether the process is an axon or dendrite.

neuropathic pain *See* neuropathy.

neuropathy Damage to nerves or abnormal activity in the nervous system.

nociceptor Mechanical, thermal or chemical receptor involved in the detection of harmful stimuli.

odorant A molecule that interacts with an olfactory receptor and elicits a smell perception.

odorant binding proteins *See* olfactory binding proteins.

odour The perception elicited by either a single odorant or a mixture of odorants.

odour response desensitization The adaptive response observed when olfactory neurons are exposed to odorants on the seconds timescale.

olfaction The sense of smell.

olfactophore (odotope) The part of a molecule that can give rise to a particular odour perception by virtue of its ability to bind to a particular combination of receptors.

olfactory/odorant binding proteins (OPBs) Proteins present in the mucus surrounding the cilia of olfactory neurons that can bind to odorant molecules; they are thought to act as carriers for odorants, transporting what are essentially hydrophobic molecules through a hydrophilic medium to the olfactory neurons.

olfactory bulb A specialized part of the brain that resides on the opposite side of the cribriform plate from the olfactory epithelium. In the olfactory bulb, the axons of the olfactory neurons synapse with mitral cells.

olfactory epithelium In humans, the 5 cm^2 patch of epithelial cells at the top of the nasal passage that houses the olfactory neurons.

olfactory mucus The secretion that coats the olfactory epithelium, containing mucopolysaccharides, immunoglobulins, enzymes and other proteins.

olfactory neuron A nerve cell in the olfactory epithelium that contains an olfactory receptor. The axon of the olfactory neuron projects to the olfactory bulb.

olfactory priming The activation by the breathing cycle of brain regions that process olfactory information.

olfactory receptor gene The sequence of chromosomal DNA that codes for an olfactory receptor protein. There are about one thousand olfactory receptor genes in the human genome, of which about a third are functional – that is, the corresponding olfactory receptor is actually synthesised.

olfactory receptor/olfactory receptor protein The protein that is embedded in the membrane of the cilium of the olfactory neuron and to which odorant molecules bind.

olfactory tubercule A structure, otherwise called the anterior perforated substance, found at the base of the olfactory tract.

operant conditioning A form of conditioning whereby the experimenter uses reinforcement or punishment to select, strengthen or weaken a behaviour pattern shown spontaneously.

overt (direction of attention) *See* overt orienting.

overt orienting Shift of attention by movement of receptors (e.g. as in eye, head or hand movements). *See also* covert orienting.

Pacinian corpuscle A type of cutaneous receptor cell that responds to heavy pressure and vibration.

papillae Structures on the tongue that house taste buds, which in turn contain the sensory taste cells. There are four types of papillae: circumvallate, filiform, foliate and fungiform.

parallel processing The simultaneous processing of several sources of stimulation. *See also* serial processing.

parietal lobe Region of the brain lying like a saddle across the top, forward of the occipital region, but behind the frontal lobes.

passive touch Being touched.

pattern theory The theory that sensations are coded for by the pattern of activity in a group of receptors.

periaqueductal gray (PAG) The region of grey matter in the brainstem that surrounds the cerebral aqueduct of the midbrain.

periglomerular cells The cells that interconnect glomeruli by forming synapses to the primary dendrites of mitral cells.

peripheral neuropathy A clinical condition where the sensory afferent fibres have been damaged or destroyed.

phantom limb pain The perception of pain from a part of the body that does not exist (usually from a limb that has been amputated).

phoneme A unit of speech sound, corresponding approximately with the sound represented by one letter.

polymodal receptor A receptor with more than one adequate stimulus, e.g. chemical, thermal and mechanical stimuli.

polymorphism The simultaneous existence of variant forms of a particular gene within a population (e.g. skin colour).

prepyriform cortex A small nucleus adjacent to the pyriform cortex with axons projecting into the pyriform area and amygdala. The area is part of the olfactory pathway and is known to show a high degree of electrophysiological coherence with the olfactory bulb. Recent evidence suggests that activity from this nucleus may project back to the bulb.

primary gustatory cortex The area of cortex responsible for the conscious perception of taste. It corresponds to Brodmann's area 43, that includes the insula and operculum and which receives information from the thalamus.

primary olfactory cortex The pyriform cortex and the orbitofrontal cortex.

primary somatosensory cortex (S1) The topographically organized area of the cerebral cortex that receives and processes sensory information from the body.

priming *See* olfactory priming.

proprioception The sense that provides information about the position and movement of individual body parts.

proprioceptor *See* proprioception.

Proust effect The phenomenon occurring when an odour prompts an emotional recollection of a personal memory.

pyriform cortex The region of five-layered cortex curled up rostrally and medially in the parahippocampal gyrus. This is primary olfactory cortex, one of three regions (the other two being the anterior olfactory nucleus and the olfactory tubercule) where the axons of the olfactory tract terminate.

ramp-and-hold stimulus An applied stimulus, the magnitude of which is increased from zero to some maximum value, and then held at that level.

reality monitoring The normal process underlying conscious awareness of the surroundings, in which current interpretations of sensory input are continually re-checked by further stimulus sampling.

receptor saturation The extent of binding of an odorant to its receptor protein depends upon the affinity between the two and the concentration of the odorant. When all of the receptors are bound to odorant then the receptor is said to be saturated and the effect of the odorant will be at its maximal value.

referred (pain) The type of pain experienced when the location of the pain is not at the site of the actual injury or pathology.

Ruffini ending A type of cutaneous sensory receptor cell that responds to self-imposed skin stretching such as that caused by moving limbs or digits.

saccades Brief rapid movements of the eye between fixation points.

sapid Having a perceptible taste or flavour, not insipid.

secondary gustatory cortex Part of the orbitofrontal cortex.

sensory neglect A condition resulting from brain damage, in which the sufferer exhibits lack of awareness of certain regions of space.

serial processing The analysis of signals from a set of items, on an item-by-item, one at a time basis. *See also* parallel processing.

shadow In the context of dichotic listening, to repeat all that one hears spoken, following very closely behind the speaker.

short-term adaptation The adaptive effect observed when olfactory neurons are exposed to very short, millisecond, pulses of odorant molecules.

signal detection theory (SDT) The theory which accounts for hits and misses in terms of neural noise and the setting of an internal threshold or criterion.

somatosensation The bodily or general senses of touch, pain, temperature, proprioception and kinesthesis.

somatosensory cortex The area of the cerebral cortex that receives and processes sensory information from the body. It includes the primary and secondary somatosensory cortices.

source attribution error A mistake in identifying the source of a piece of information, as when a hypnotized person is taken through an imagined experience, which they subsequently believe to have happened in reality.

special senses The senses of vision, hearing, balance, taste and smell for which the receptors are located in specialized sensory organs.

spinothalamic tract (STT) The neural pathway followed by axons through the spinal cord to the level of the thalamus and beyond.

stratum corneum The relatively hard surface of the skin formed by a layer of dead epithelial cells.

sub-cortical Regions of the brain beneath the cortex.

subliminal Term used to describe a stimulus which has insufficient intensity to produce conscious awareness (although other measures may suggest that it was nevertheless detected).

substantia gelatinosa (SG) A region of the dorsal horn of the spinal cord (also known as Rexed's lamina II). There are many interneurons that are believed to be important in modulating pain sensations in this area.

superior turbinate One of the three structures that arise from the lateral wall of the nasal cavity, the function of which is to create turbulent air flow and to clean, moisten and warm the inhaled air. The superior turbinate is a projection of the ethmoid bone, and is the uppermost and smallest of the three.

suppression The decrease in intensity of a particular taste quality of one tastant due to the presence of another tastant.

synergy An increase in intensity of a particular taste quality that is brought about by the presence of another tastant, the intensity being more than the sum of the two individual tastants.

tachistoscope A device for presenting visual stimuli for brief, controlled lengths of time, e.g. as when words are flashed on a screen for a fraction of a second.

taction The mechanical aspects of somatosensation.

tastant A molecule that interacts with a taste cell and elicits a gustatory perception.

taste bud A group of cells, resembling a bud, that contains sensory taste cells which respond to several taste qualities, namely salt, sour, sweet, bitter and umami. The sensory taste cells are innervated by either the chorda tympani or the glossopharyngeal nerve.

TENS (transcutaneous electrical nerve stimulation) A technique for pain reduction using electrical stimuli directed through the skin. The stimuli preferentially excite myelinated nerve fibres and may activate inhibitory neurons or stimulate the release of endorphins within the brain.

thalamus The part of the forebrain, formed from a collection of about 30 different nuclei, that relays sensory information to the cerebral cortex.

tight junction The junction formed between two cells that is permeable only to small ions and molecules.

topographic map The orderly representation of the spatial arrangement of sensory receptors through successive stages of neural processing along the ascending pathway to the cortex. This representation therefore preserves the spatial relationships of the activation pattern of the receptors.

trigeminal neuralgia Episodes of intense facial pain caused by a disorder of the trigeminal nerve.

trigeminal system The neural system that senses mechanical and other somatosensory stimuli on the face.

umami One of the main taste categories; a Japanese word meaning 'delicious savoury taste'. It is common in oriental foods and is the taste of amino acids, the most well known of which is monosodium glutamate. It is triggered by molecules binding to a gustatory glutamate G-coupled receptor.

unimodal Operating in one modality, i.e. associated with one sense only.

utricular cavity The part of the vestibule, in the inner ear, with which the semicircular canals communicate.

vagus (Xth cranial nerve) A parasympathetic nerve that supplies structures in the head, neck, thorax, and abdomen. In the head it is cranial nerve X and innervates taste buds in the lining of the mouth.

vallate papillae *See* circumvallate papillae.

vestibular labyrinth The group of structures which constitute the balance organs.

virtual reality headset A helmet-like device, with built-in displays for each eye, and often with stereo headphones, which presents computer-generated stimuli so that the wearer feels as if in a real world.

viscera The interior organs of the body, particularly in the abdomen (e.g. the intestines).

visual pursuit reflex The tracking action of the eyes, produced when an object being observed is moving across the field of view.

volatile In a chemical context, volatile compounds are those that have high vapour pressures, that is, they are easily vaporized.

vomeronasal organ A chemosensory organ, also known as Jacobson's organ, part of an accessory olfactory system specialized for the detection of pheromones. Pheromones are chemicals which mediate sexual/mating behaviour (but that may have no perceivable odour). The vomeronasal organ is located in the nasal septum or roof of the mouth in vertebrates. In humans, the organ is present but appears to be vestigial (i.e. not functional).

wind-up Also known as central sensitization, it is the increased pain susceptibility brought about by the increasing response of dorsal horn neurons to a steady, prolonged input from nociceptors.

Index

Entries and page numbers in **bold type** refer to key words which are printed in **bold** in the text and which are defined in the Glossary.

A

Aδ fibres 41
ACE *see* angiotensin-converting enzyme
acesulfame 122, 126
  and aspartame 141
acetic acid *64*, 133–4, *135*, 136–7
acids
  and pH scale 136
  sour taste of 116, 133–8
active touch 21, 22
acute inflammatory response 33
  see also inflammation
adaptation (in olfactory system) 89–93
  long-lasting 92, *93*
  odour response desensitization 91–2
  short-term 89–91
adaptation of mechanoreceptors 11–12, 13, 18
addiction 44
  fear of 32
adenylyl cyclase 73, 91, *92*
adequate stimulus 5
  skin receptors *8*
aircraft, take-off from aircraft carrier 181–2
alcohols 76, 77, 95, 96, *97*
  see also ethanol
aldehydes 69–70, 81–2, *97*
  see also benzaldehyde
allicin *64*
allodynia 38, 39–40
amiloride 129, 139–40, 151
Amoore, John, theory of odorants 71, 72
amputees, phantom limb pain in 46–8
amygdala 62, 146, 159, 160
analgesic substances **32**
anandamide 42
anethole *64*
angiotensin-converting enzyme (ACE), inhibitors 34
anisole *100*, *101*
arachidonic acid cascade 34, 36
arginine 141
Aristotle's illusion 21
arthritis 37
ascorbic acid 133
aspartame 122, 123, 143
  and acesulfame 141
  detection in mixtures 143
aspartic acid 131
aspirin 37, 110, 114, *115*
attention
  covert and **overt 180**
  failures of 183–4
  and hearing 177
  and vision 173–5

B

balance 28–9
Balint's syndrome 183
benzaldehyde *64*, 157–8
  compared with hydrogen cyanide 63, 65, 105
  detection in mixtures 157–8
benzyl acetate *93*, *94*
Bitrex® *see* denatonium benzoate
bitter tasting materials 110, 114, 115, 117–21
body image 47
boiling temperatures 62–3
bradykinin 34, *35*, 36, *38*
brain
  frontal lobes 185
  integration of sensory information 171
  odorant-responsive regions of *61*, 62, 73, 85–6, *89*
  parietal lobes 183–4, 185
  taste-responsive regions of 145–6
breastfeeding *23*
Brillat-Savarin, Jean-Anthelme 109, 121, 155
bruising 33
butanoic acid 133
2-butanol *98*

C

C fibres 41, 44
C-terminus 67, *68*
caffeine 114, *115*, 117
calcitonin gene-related peptide 33, *34*
calcium levels
  in olfactory neurons 74–5, 87, *88*, 90–1, 92
  in taste receptors 119, 125
calmodulin 91
CaMKII kinase 91–2
camphor *98*
capsaicin 42
carbonated water 110, 138
carbonic anhydrase 138
carboxylic acids *76*, 77, *85*, 96, *97*
carvone 80–1, 97, *98*, 100
celecoxib 37
chemical senses 59
chemotopic maps (taste) **147**, *148*
chiral recognition 96, **98**–100
chorda tympani 111, *112*, 144, *145*, 149, *150*
  and sour taste 134, *135*, 137, 138
  and umami taste 129
chronic pain 48
cineole 90, *91*, 92
cingulate cortex 48
circumvallate papillae 111, *112*, 119, 130, 147, *148*
citric acid 133
  and sucrose 142
citronellol *98*
coffee 155
combinatorial odour code 75–9, 80, 81, 89
  equation for 77
complement *34*
configurational hypothesis of olfaction 102
congruence 158, 160
contact lenses 10
cornea, mechanoreceptors in 9, 10
cortical processing 19, 28, 183
covalent bonds 64, 65
  polarization of 65–6
covert orienting 180
cranial nerve IX *see* glossopharyngeal nerve
cranial nerve VII *see* chorda tympani
crossmodal interactions 157, 159
cyclamate *see* sodium cyclamate

cyclic AMP (cAMP) 73, *74*, 91–2
cycloheximide 120
cyclo-oxygenases (COX-1, COX-2) ***36***, 37

D

denatonium benzoate 114, *115*, 119
depolarization of cells 73–4
desensitization *see* odour response desensitization
dichotic listening 177
dorsal root ganglia 16, 37
double bonds 4
double dissociation (stimulus/sensation) **31**, *32*
dulcin 114, *115*

E

EC_{50} 88
endogenous attention shifts **180**, 184
enhancement (tastant interactions) **141**
enkephalins 43
epidural analgesia 46
esters, boiling temperatures 62–3
ethanol (alcohol), effects of 182
ethyl glycidate *100*, *101*
exogenous attention shifts **180**, 183
extrafusal muscle fibres *27*

F

fast-adapting fibres 18
feature integration 174–5
feedback inhibition 91
fenchone *98*
field receptors *8*
filiform papillae 111, *112*
flavour 109, 155–61
fMRI *see* functional magnetic resonance imaging
foliate papillae 111, *112*, 119, 130, 147, *148*
forced-choice triangular test 94–5, 96, *97*
free nerve endings *7*, *8*, 9, 17
  in muscle spindles 26
  nociceptors 37
frontal operculum *145*, 146
fructose 121–2, 126
functional magnetic resonance imaging (fMRI)
  odorant response 85
  taste response 145
fungiform papillae 111, *112*, 118, 119, 137, *148*
fura-2 74

G

G-protein-coupled receptors 73, 74, *91*, *92*, 119, 125, 130
gate-control theory of pain 32, 43–44
general senses 5
genes
  for olfactory receptors 72, 75, 78, 87
  for taste receptors 119
genome 72
geraniol 93, *94*
glomeruli 72, 73, **79**–84, 89
glossopharyngeal nerve 111, *112*, 144, *145*, 149, *150*
glucose 121–2
  brain regions activated by *145*, 146
glutamate receptors 129–32
glutamic acid 115, *116*
glycerol (glycerine) 109, 113, *114*, 122
GMP *see* guanosine monophosphate
Golgi tendon organs 25, *26*
granule cells 83
grieving 23
guaiacol *100*, *101*
guanosine monophosphate (GMP) 115, *116*, 131
gustation 109
gustatory neurons 149, 160–1
gustatory system 111–13
gymnemic acid 142

H

Hageman factor *35*
hair follicle receptors *see* root hair plexus of follicle
head movements 28–9
hearing, and attention 177
hedonic perception 159, 161
hefting 20
hippocampus *61*
histamine *34*
histological studies **9**
hydrochloric acid 134, *135*, 137, 150–1
hydrogen bonding 66
hydrogen cyanide *64*
  compared with benzaldehyde 63, 65, 105
hyperalgesia 38, 39–40
hypnosis 185
hypothalamus 62

I

ibuprofen 37, 98
illusions
  Aristotle's 21
  of climbing aircraft 181–2
  following heavy drinking 182
imagined perceptions 185
IMAX® cinemas 181
immune system, and inflammatory response 33, 34–7
indole 87, 93, *94*
inflammation 33–7, 38
inosine monophosphate 131
insula *145*, 146, 159
integration, failures of 183–4
intrafusal muscles 26, *27*
intrinsic signal imaging 80–1
ion channels 17, 42, 43
  response to odorants 73–4, 91
  response to salt taste 139–40
ischemia 43

J

jasmine fragrance 93, *94*
jasmine lactone 93, *94*
jasmone 93, *94*
joint receptors 25, 28

K

kallidin 34, *35*
kallikrein *35*
keratoconus 10
ketones 95, 96, *97*
kinesthesis 25
kinins 34–6

L

labelled-line theory 75
  of odour coding 75, 77
  of taste coding 121, 146
library experiment 23
limonene *64*, 66, 97, *98*, 100
  compared with menthol 65
lithium chloride 116
London forces 66
long-lasting adaptation 92, *93*
lysine residue
  in odorant binding

M

McGurk effect 179
D-mannose isomers 114–15
masking (tastant interactions) **142**, 143–4
mast cells 33, *34*, *38*
mechanoreceptors 5, *6*
adaptation of 11–12, 13, 18
see also proprioceptors; tactile mechanoreceptors
medial lemniscus *112*, *145*
Meissner's corpuscles *7*, *8*, *9*, 10, 11, 12, 14
memory 184, 187
Menière's disease 181
menthol *64*, *98*
compared with limonene 65
Merkel's discs *7*, *8*, *9*, 10, 12, 13, 15
methyl jasmonate 93, *94*
methyl salicylate *100*, *101*, 102
2-methyl-3-furyl disulfide 155–6, 158
mirror image molecules *see* chiral recognition
mitral cells 79–80, 82–4
mobile telephones, used in cars 180
molecular interactions 65–6, 87, 113
molecular receptive range 69–**70**, 81–2
monosodium aspartate (MSA) 131
monosodium glutamate (MSG) 110, 129, 155–6
and benzaldehyde 157–8
and 2-methyl-3-furyl disulfide 155–6, 158
and ribonucleotides 141
MSA *see* monosodium aspartate
MSG *see* monosodium glutamate
muscle spindles 25, 26, *27*, 28

N

N-terminus 67, *68*, 120, 125, 130, 132
neural coding 175
neurons, somatosensory 5, 17
neuropathic pain **31**, 48
neurotransmitter receptors 130
nociceptors 5–6, 37–41, 43
thermo- 39, 41
nona-2,6-dienal 94
non-steroidal anti-inflammatory drugs (NSAIDs) 37
NST *see* solitary tract, nucleus

O

octanal 68, 69–70, 72, *81*
odorants 62–6
binding to olfactory receptors 69–70, 87–9
chiral recognition 96, 98–100
interactions with tastants 157, 160
odour 60, 62
odour coding
combinatorial 75–9, 80, 81, 89
labelled-line theory 75, 77
odour detection 78–9
odour discrimination 78–9, 93–105
odour perception, effect of concentration on 87–9
odour response desensitization 91–2
olfaction 61–2
configurational hypothesis of 102
olfactophores 72
olfactory bulb *61*, **62**, 80–1, *89*
olfactory epithelium 61–2, 72
olfactory neurons 62, 67, 72, *73*, *160*, 161
calcium concentrations in 74–5, 87, *88*, 90–1, 92
olfactory receptors *61*, 66, **67**–9, 79
chiral recognition by 96, 98–100
distribution of 71–2
genes for 72, 75, 78, 87
molecular receptive range of 69–70
and odour coding 75–9
saturation *see* receptor saturation
transduction of information in 73–5
olfactory system *61*, 116
dynamic picture of 78
olfactory tubercule *61*
opiates 43–4
opioids 32
orbitofrontal cortex *145*, 146, *147*, 149, *150*, 159, 160, 161
otoliths 181
overt orienting **180**

P

Pacinian corpuscles *7*, *8*, *9*, 10, 13, 14, 17
pain 5–6, 31–49
control of 31–2, 44–8
gate-control theory of 32
palate, taste buds on 111, 119, 147
papillae 111, *112*, *148*
parallel processing 173–4
pentyl acetate 80–1, 89
perception(s)
imagined and real 185
of pain 6
of touch 19–23
perfumery compounds 71–2
periglomerular cells 83
PET *see* positron emission tomography
pH scale 136
phantom limb pain 46–8
phenylthiocarbamide (PTC) 114, *115*
ability to taste 118
α-pinene 97, *98*, 100
point localization thresholds *15*, 16
polar attractions 65
polar substances 113
polarization of bonds 65–6
polymodal receptors 5, 37
response to heat stimuli 39, *40*
VR-1 receptors in 42
polymorphism 118
for bitter taste 120
positron emission tomography (PET)
odour response 85, 86
taste response 146, *147*
posteromedial ventral nucleus of thalamus (PVN) *112*, *145*
potassium chloride 116, 149–50, 151, *152*
pressure, response to 10–11, 14–16
primary gustatory cortex 146
primary olfactory cortex 79, 84
priming 187
propanal 81
propanoic acid 81
proprioception 5, 25–9
loss of input from 29
proprioceptors *6*, 25–7
6-propylthiouracil (PROP), ability to taste 118–19
prostaglandins 36, *38*
Proust, Marcel 59
pseudogenes 75
psychophysical studies 9, 13–14, 155–6
PTC *see* phenylthiocarbamide
PVN *see* posteromedial ventral nucleus of thalamus

Q

quinine 110, 114, *115*, 117
rejection threshold 117
and sodium chloride 142

R

ramp-and-hold stimulus **12**
rapidly adapting fibres 18
rapidly adapting receptors 11–14
reading, absorption in 183, 185
receptor saturation 88
receptors
  active in somatosensation 6
  see also mechanoreceptors; nociceptors; proprioceptors; thermoreceptors
referred pain 44, *45*
ribonucleotides 115, 131–2
  and MSG 141
rofecoxib 37
root hair plexus of follicle *7*, *8*, 11
rose oxide, odorant *98*
Ruffini's endings *7*, *8*, *9*, 12, 13

S

saccharin 109, 113, 114, 122, 125
  and benzaldehyde 157, 158
salt 109
salty tasting materials 116, 139–41
sea-sickness 182
secondary gustatory cortex 146
selective attention 179–80
sensory neglect 184
serial processing 173, 174
shadowing 177
short-term adaptation 89–91
signal detection theory 187
silent nociceptors 37, 44
skin receptors *7*, *8*
  see also tactile mechanoreceptors; thermoreceptors
skin stretch, detection 13
slowly adapting fibres 18
slowly adapting receptors 11–12
smell, sense of 59, 61–2, *156*
  configurational hypothesis of 102
  see also flavour
sodium chloride 116, 139–40, 141, 149, 150–1, *152*
  brain regions activated by *145*, 146
  detection in mixtures 142, 143–4
sodium cyclamate 122
sodium saccharin *see* saccharin
solitary tract 149, 150–1, 152
  nucleus (NST) *112*, 144
somatosensation 5–6, 28
somatosensory coding 17–18
somatosensory cortex 46, 47
somatosensory neurons 5, 17
sour tasting materials 110, 116, 133–8
special senses 5
spinal analgesia 46
staphylococcal infections 33
stimuli
  meaning of 175
  related to sensation 31, *32*, 41
  unattended 183
streptococcal infections 33
stroke victims 183
strychnine 117
sub-cortical processing **183**
subliminal advertising **175**
substance P 33, *34*, *38*, 43
sucrose 109, 121–2
  and citric acid 142
  enhancement/reduction of sweetness of 158–9
sugar *see* sucrose
supertasters 118
suppression (tastant interactions) **142**
sweet tasting materials 109, 113, 114, 121–8, 141
sweetness, measuring 124
synergy (tastant interactions) **141**

T

tachistoscope, experiments using **174**–5
tactile mechanoreceptors 7–17
taction (touch) **7**–18
tastants 113–16
  binding to receptors 120, 123–8
  detection in mixtures 141–4, 246–7
  interactions with odorants 157, 160
taste
  sense of 59–60, 109–10, *156*
  see also flavour
taste buds 111
taste cells 111, 125, 144, 147, 149
taste coding 121, 146–52
taste interactions 141–4
taste receptors
  bitter 118–21, 130, 147
  sour 134, 137–8
  sweet 123, 124–8, 130, 147
  umami 130–2
temperature, sense of *see* thermal detection
α-terpineol *99*
thalamus *61*, 111
thalidomide 99
thaumatin 113, *114*
thermal detection 6, 9
thermoreceptors *6*, 20
  response to heat stimuli 39
threshold of detection, pressure 14–16
tight junctions 139, 140
tissue damage 33–7, 44
  detection 9
  pain of 31, *32*
tolerance of drugs 44
topographic maps (taste) **147**, *148*
touch 5
  perception of 19–23
  social aspects of 23
  taxonomy of 19–20
touch receptors *6*
  see also tactile mechanoreceptors
T1R-1 receptors 125, 128
T1R-2 receptors 125–8
T1R-3 receptors 125–6, 127–8
T2R-1 receptors 119–20
transduction
  of noxious stimuli 41–3
  of somatosensory stimuli 17
transmembrane (TM) domains 68–9, 120
triple bonds 64

U

umami 110
  tastants 115, 116, 128–32, 141, 155–7
urea, and sodium chloride 142

V

vanillin 85, 87–9
vestibular system 28
  and vision 181–2
vibratory stimuli
  perceptions elicited *9*
  response to 10, 13–14
viscera 5
visceral pain 44
vision
  and attention 173–5
  contribution to balance and movement 28–9
  and hearing 179–80
  and vestibular system 181–2
visual information, integrating 171
visual neglect 187
volatile substances **62**–3
VR-1 receptors 42

W

wind-up 45–6, 47